STUDENT WORKBOOK
VOLUME 1

P H Y S I C S

FOR SCIENTISTS AND ENGINEERS SECOND EDITION

A STRATEGIC APPROACH

Randall D. Knight
California Polytechnic State University
San Luis Obispo

PEARSON
Addison
Wesley

San Francisco   Boston   New York
Capetown   Hong Kong   London   Madrid   Mexico City
Montreal   Munich   Paris   Singapore   Sydney   Tokyo   Toronto

Publisher: Adam Black, Ph.D.
Development Manager: Michael Gillespie
Development Editor: Alice Houston, Ph.D.
Project Editor: Martha Steele
Assistant Editor: Grace Joo
Media Producer: Deb Greco
Sr. Administrative Assistant: Cathy Glenn
Director of Marketing: Christy Lawrence
Executive Marketing Manager: Scott Dustan
Sr. Market Development Manager: Josh Frost
Market Development Associate: Jessica Lyons
Managing Editor: Corinne Benson
Production Supervisor: Nancy Tabor
Production Service: WestWords PMG
Illustrations: Precision Graphics
Text Design: Seventeenth Street Studios and WestWords PMG
Cover Design: Yvo Riezebos Design and Seventeenth Street Studios
Manufacturing Manager: Pam Augspurger
Text and Cover Printer: Edwards Brothers
Cover Image: Composite illustration by Yvo Riezebos Design; photo of spring by Bill Frymire/Masterfile

ISBN-13: 978-0-321-51626-8
ISBN-10: 0-321-51626-5

1 2 3 4 5 6 7 8 9 10—EB—10 09 08 07
www.aw-bc.com

Table of Contents

Preface

Learning physics, just as learning any skill, requires regular practice of the basic techniques. That is what this *Student Workbook* is all about. The workbook consists of exercises that give you an opportunity to practice the ideas and techniques presented in the textbook and in class. These exercises are intended to be done on a daily basis, right after the topics have been discussed in class and are still fresh in your mind.

You will find that the exercises are nearly all *qualitative* rather than *quantitative*. They ask you to draw pictures, interpret graphs, use ratios, write short explanations, or provide other answers that do not involve significant calculations. The purpose of these exercises is to help you develop the basic thinking tools you'll later need for quantitative problem solving. Successful completion of the workbook exercises will prepare you to tackle the more quantitative end-of-chapter homework problems in the textbook. It is highly recommended that you do the workbook exercises *before* starting the end-of-chapter problems.

You will find that the exercises in this workbook are keyed to specific sections of the textbook in order to let you practice the new ideas introduced in that section. You should keep the text beside you as you work and refer to it often. You will usually find Tactics Boxes, figures, or examples in the textbook that are directly relevant to the exercises. When asked to draw figures or diagrams, you should attempt to draw them so that they look much like the figures and diagrams in the textbook.

Because the exercises go with specific sections of the text, you should answer them on the basis of information presented in *just* that section (and prior sections). You may have learned new ideas in Section 7 of a chapter, but you should not use those ideas when answering questions from Section 4. There will be ample opportunity in the Section 7 exercises to use that information there.

You will need a few "tools" to complete the exercises. Many of the exercises will ask you to *color code* your answers by drawing some items in black, others in red, and yet others in blue. You need to purchase a few colored pencils to do this. The author highly recommends that you work in pencil, rather than ink, so that you can easily erase. Few people produce work so free from errors that they can work in ink! In addition, you'll find that a small, easily carried six-inch ruler will come in handy for drawings and graphs.

As you work your way through the textbook and this workbook, you will find that physics is a way of *thinking* about how the world works and why things happen as they do. We will be interested primarily in finding relationships and seeking explanations, only secondarily in computing numerical answers. In many ways, the thinking tools developed in this workbook are what the course is all about. If you take the time to do these exercises regularly and to review the answers, in whatever form your instructor provides them, you will be well on your way to success in physics.

To the instructor: The exercises in this workbook can be used in many ways. You can have students work on some exercises in class as part of an active-learning strategy. Or you can do the same in recitation sections or laboratories. This approach allows you to discuss the answers immediately, to answer student questions, and to improvise follow-up exercises when needed. Having the students work in small groups (2 to 4 students) is highly recommended.

Alternatively, the exercises can be assigned as homework. The pages are perforated for easy tear-out, and the page breaks are in logical places so that you can assign the sections of a chapter that you would likely cover in one day of class. Exercises should be assigned immediately after presenting the relevant information in class and should be due at the beginning of the next class. Collecting them at the beginning of class, then going over two or three that are likely to cause difficulty, is an effective means of quickly reviewing major concepts from the previous class and launching a new discussion.

If the exercisees are used as homework, it is *essential* for students to receive *prompt* feedback. Ideally this would occur by having the exercises graded, with written comments, and returned at the next class meeting. Posting the answers on a course website also works. Lack of prompt feedback can negate much of the value of these exercises. Placing similar qualitative/ graphical questions on quizzes and exams, and telling students at the beginning of the term that you will do so, encourages students to take the exercises seriously and to check the answers.

The author has been successful with assigning *all* exercises in the workbook as homework, collecting and grading them every day through Chapter 4, then collecting and grading them on about one-third of subsequent days on a random basis. Student feedback from end-of-term questionnaires reveals three prevalent attitudes toward the workbook exercises:

i.   They think it is an unreasonable amount of work.

ii. They agree that the assignments force them to keep up and not get behind.

iii. They recognize, by the end of the term, that the workbook is a valuable learning tool.

However you choose to use these exercises, they will significantly strengthen your students' conceptual understanding of physics.

Following the workbook exercises are optional Dynamics Worksheets, Momentum Worksheets, and Energy Worksheets for use with end-of-chapter problems in Parts I and II of the textbook. Their use is recommended to help students acquire good problem-solving habits early in the course. End-of-chapter problems marked with the ✐ icon are intended to be done on worksheets.

Answers to all workbook exercises are provided as pdf files on the *Media Manager*. The author gratefully acknowledges the careful work of answer writers Professor James H. Andrews of Youngstown State University and Rebecca Sobinovsky.

Acknowledgments: Many thanks to Martha Steele at Addison-Wesley and to Jared Sterzer at WestWords PMG for handling the logistics and production of the *Student Workbook.*

1 Concepts of Motion

1.1 Motion Diagrams

1.2 The Particle Model

Exercises 1–5: Draw a motion diagram for each motion described below.
- Use the particle model to represent the object as a particle.
- Six to eight dots are appropriate for most motion diagrams.
- Number the positions in order, as shown in Figure 1.4 in the text.
- Be neat and accurate!

1. A car accelerates forward from a stop sign. It eventually reaches a steady speed of 45 mph.

2. An elevator starts from rest at the 100th floor of the Empire State Building and descends, with no stops, until coming to rest on the ground floor. (Draw this one *vertically* since the motion is vertical.)

3. A skier starts *from rest* at the top of a 30° snow-covered slope and steadily speeds up as she skies to the bottom. (Orient your diagram as seen from the *side*. Label the 30° angle.)

4. The space shuttle orbits the earth in a circular orbit, completing one revolution in 90 minutes.

5. Bob throws a ball at an upward 45° angle from a third-story balcony. The ball lands on the ground below.

Exercises 6–9: For each motion diagram, write a short description of the motion of an object that will match the diagram. Your descriptions should name *specific* objects and be phrased similarly to the descriptions of Exercises 1 to 5. Note the axis labels on Exercises 8 and 9.

6.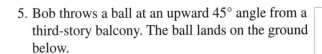

8.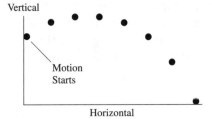

7.

9.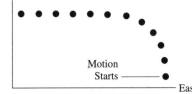

1.3   Position and Time

10. The figure below shows the location of an object at three successive instants of time.

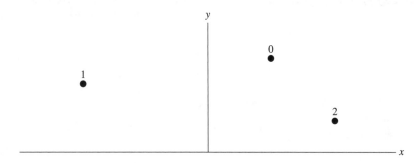

   a. Use a **red** pencil to draw and label on the figure the three position vectors $\vec{r}_0$, $\vec{r}_1$, and $\vec{r}_2$ at times 0, 1, and 2.
   b. Use a **blue** or **green** pencil to draw a possible trajectory from 0 to 1 to 2.
   c. Use a **black** pencil to draw the displacement vector $\Delta\vec{r}$ from the initial to the final position.

11. In Exercise 10, is the object's displacement equal to the distance the object travels? Explain.

12. Redraw your motion diagrams from Exercises 1 to 5 in the space below. Then add and label the displacement vectors $\Delta\vec{r}$ on each diagram.

1.4 Velocity

13. The figure below shows the positions of a moving object in three successive frames of film. Draw and label the velocity vector $\vec{v}_0$ for the motion from 0 to 1 and the vector $\vec{v}_1$ for the motion from 1 to 2.

Exercises 14–20: Draw a motion diagram for each motion described below.
- Use the particle model.
- Show and label the *velocity* vectors.

14. A rocket-powered car on a test track accelerates from rest to a high speed, then coasts at constant speed after running out of fuel. Draw a dashed line across your diagram to indicate the point at which the car runs out of fuel.

15. Galileo drops a ball from the Leaning Tower of Pisa. Consider the ball's motion from the moment it leaves his hand until a microsecond before it hits the ground. Your diagram should be vertical.

16. An elevator starts from rest at the ground floor. It accelerates upward for a short time, then moves with constant speed, and finally brakes to a halt at the tenth floor. Draw dashed lines across your diagram to indicate where the acceleration stops and where the braking begins. You'll need 10 or 12 points to indicate the motion clearly.

17. A bowling ball being returned from the pin area to the bowler starts out rolling at a constant speed. It then goes up a ramp and exits onto a level section at very low speed. You'll need 10 or 12 points to indicate the motion clearly.

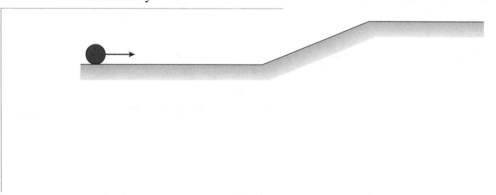

18. A track star runs once around a running track at constant speed. The track has straight sides and semicircular ends. Use a bird's-eye view looking down on the track. Use about 20 points for your motion diagram.

19. A car is parked on a hill. The brakes fail, and the car rolls down the hill with an ever-increasing speed. At the bottom of the hill it runs into a thick hedge and gently comes to a halt.

20. Andy is standing on the street. Bob is standing on the second-floor balcony of their apartment, about 30 feet back from the street. Andy throws a baseball to Bob. Consider the ball's motion from the moment it leaves Andy's hand until a microsecond before Bob catches it.

1.5   Linear Acceleration

Note: Beginning with this section, and for future motion diagrams, you will "color code" the vectors. Draw velocity vectors **black** and acceleration vectors **red**.

Exercises 21–24: The figures below show an object's position in three successive frames of film. The object is moving in the direction $0 \rightarrow 1 \rightarrow 2$. For each diagram:

- Draw and label the initial and final velocity vectors $\vec{v}_0$ and $\vec{v}_1$. Use **black**.
- Use the steps of Tactics Box 1.3 to find the change in velocity $\Delta \vec{v}$.
- Draw and label $\vec{a}$ at the proper location on the motion diagram. Use **red**.
- Determine whether the object is speeding up, slowing down, or moving at a constant speed. Write your answer beside the diagram.

21.

22.

23.

24.

Exercises 25–29: Draw a complete motion diagram for each of the following.
- Draw and label the velocity vectors $\vec{v}$. Use **black**.
- Draw and label the acceleration vectors $\vec{a}$. Use **red**.

25. Galileo drops a ball from the Leaning Tower of Pisa. Consider its motion from the moment it leaves his hand until a microsecond before it hits the ground.

26. Trish is driving her car at a steady 30 mph when a small furry creature runs into the road in front of her. She hits the brakes and skids to a stop. Show her motion from 2 seconds before she starts braking until she comes to a complete stop.

27. A ball rolls up a smooth board tilted at a 30° angle. Then it rolls back to its starting position.

28. A bowling ball being returned from the pin area to the bowler rolls at a constant speed, then up a ramp, and finally exits onto a level section at very low speed.

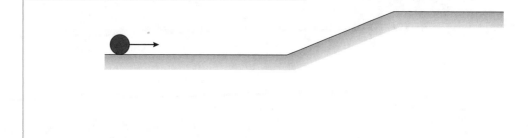

29. Two sprinters, Cynthia and Diane, start side by side. Diane has run only 80 m when Cynthia crosses the finish line of the 100 m dash.

1.6   Motion in One Dimension

1.7   Solving Problems in Physics

30. The four motion diagrams below show an initial point 0 and a final point 1. A pictorial representation would define the five symbols: x_0, x_1, v_{0x}, v_{1x}, and a_x for horizontal motion and equivalent symbols with y for vertical motion. Determine whether each of these quantities is positive, negative, or zero. Give your answer by writing +, −, or 0 in the table below.

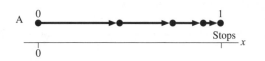

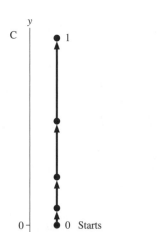

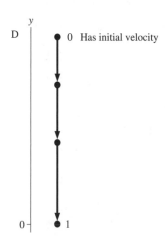

	A	B	C	D
x_0 or y_0				
x_1 or y_1				
v_{0x} or v_{0y}				
v_{1x} or v_{1x}				
a_x or a_y				

31. The three symbols x, v_x, and a_x have eight possible combinations of *signs*. For example, one combination is $(x, v_x, a_x) = (+, -, +)$.

    a. List all eight combinations of signs for x, v_x, a_x.

    1. _____   5. _____

    2. _____   6. _____

    3. _____   7. _____

    4. _____   8. _____

b. For each of the eight combinations of signs you identified in part a:
- Draw a four-dot motion diagram of an object that has these signs for x, v_x, and a_x.
- Draw the diagram *above* the axis whose number corresponds to part a.
- Use **black** and **red** for your $\vec{v}$ and $\vec{a}$ vectors. Be sure to label the vectors.

1. ——————————————————————— x
   0

2. ——————————————————————— x
   0

3. ——————————————————————— x
   0

4. ——————————————————————— x
   0

5. ——————————————————————— x
   0

6. ——————————————————————— x
   0

7. ——————————————————————— x
   0

8. ——————————————————————— x
   0

32. Sketch position-versus-time graphs for the following motions. Include a numerical scale on both axes with units that are *reasonable* for this motion. Some numerical information is given in the problem, but for other quantities make reasonable estimates.

    Note: A *sketched* graph simply means hand-drawn, rather than carefully measured and laid out with a ruler. But a sketch should still be neat and as accurate as is feasible by hand. It also should include labeled axes and, if appropriate, tick-marks and numerical scales along the axes.

    a. A student walks to the bus stop, waits for the bus, then rides to campus. Assume that all the motion is along a straight street.

    b. A student walks slowly to the bus stop, realizes he forgot his paper that is due, and *quickly* walks home to get it.

    c. The quarterback drops back 10 yards from the line of scrimmage, then throws a pass 20 yards to the tight end, who catches it and sprints 20 yards to the goal. Draw your graph for the *football*. Think carefully about what the slopes of the lines should be.

33. Interpret the following position-versus-time graphs by writing a very short "story" of what is happening. Be creative! Have characters and situations! Simply saying that "a car moves 100 meters to the right" doesn't qualify as a story. Your stories should make *specific reference* to information you obtain from the graphs, such as distances moved or time elapsed.

a. Moving car

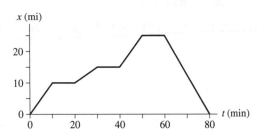

b. Submarine

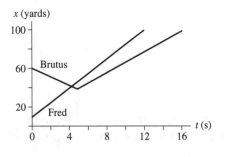

c. Two football players

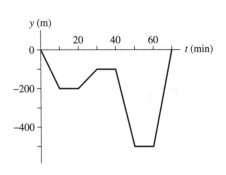

34. Can you give an interpretation to this position-versus-time graph? If so, then do so. If not, why not?

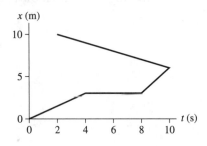

© 2008 by Pearson Education, Inc., publishing as Pearson Addison-Wesley.

1.8   Units and Significant Figures

35. Convert the following to SI units. Work across the line and show all steps in the conversion.

   a. 9.12 μs ×

   b. 3.42 km ×

   c. 44 cm/ms ×

   d. 80 km/hr ×

   e. 60 mph ×

   f. 8 in ×

   g. 14 in^2 ×

   h. 250 cm^3 ×

   Note: Think carefully about g and h. A picture may help.

36. Use Table 1.5 to assess whether or not the following statements are *reasonable*.

   a. Joe is 180 cm tall.

   b. I rode my bike to campus at a speed of 50 m/s.

   c. A skier reaches the bottom of the hill going 25 m/s.

d. I can throw a ball a distance of 2 km.

e. I can throw a ball at a speed of 50 km/hr.

37. Justify the assertion that 1 m/s ≈ 2 mph by *exactly* converting 1 m/s to English units. By what percentage is this rough conversion in error?

38. How many significant figures does each of the following numbers have?

a. 6.21 _____

b. 62.1 _____

c. 6210 _____

d. 6210.0 _____

e. 0.0621 _____

f. 0.620 _____

g. 0.62 _____

h. .62 _____

i. 1.0621 _____

j. 6.21×10^3 _____

k. 6.21×10^{-3} _____

l. 62.1×10^3 _____

39. Compute the following numbers, applying the significant figure standards adopted for this text.

a. $33.3 \times 25.4 =$ _____

b. $33.3 - 25.4 =$ _____

c. $33.3 \div 45.1 =$ _____

d. $33.3 \times 45.1 =$ _____

e. $2.345 \times 3.321 =$ _____

f. $(4.32 \times 1.23) - 5.1 =$ _____

g. $33.3^2 =$ _____

h. $\sqrt{33.3} =$ _____

2 Kinematics in One Dimension

2.1 Uniform Motion

1. Sketch position-versus-time graphs for the following motions. Include appropriate numerical scales along both axes. A small amount of computation may be necessary.

   a. A parachutist opens her parachute at an altitude of 1500 m. She then descends slowly to earth at a steady speed of 5 m/s. Start your graph as her parachute opens.

   b. Trucker Bob starts the day 120 miles west of Denver. He drives east for 3 hours at a steady 60 miles/hour before stopping for his coffee break. Let Denver be located at $x = 0$ mi and assume that the x-axis points to the east.

   c. Quarterback Bill throws the ball to the right at a speed of 15 m/s. It is intercepted 45 m away by Carlos, who is running to the left at 7.5 m/s. Carlos carries the ball 60 m to score. Let $x = 0$ m be the point where Bill throws the ball. Draw the graph for the *football*.

2. The figure shows a position-versus-time graph for the motion of objects A and B that are moving along the same axis.

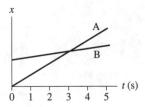

   a. At the instant $t = 1$ s, is the speed of A greater than, less than, or equal to the speed of B? Explain.

   b. Do objects A and B ever have the *same* speed? If so, at what time or times? Explain.

3. Interpret the following position-versus-time graphs by writing a short "story" about what is happening. Your stories should make specific references to the *speeds* of the moving objects, which you can determine from the graphs. Assume that the motion takes place along a horizontal line.

   a.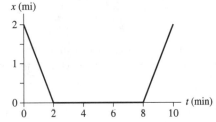

   b.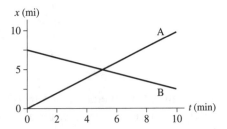

   c.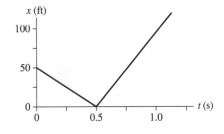

2.2 Instantaneous Velocity

4. Draw both a position-versus-time graph *and* a velocity-versus-time graph for an object at rest at $x = 1$ m.

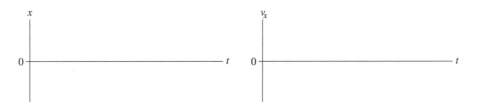

5. The figure shows the position-versus-time graphs for two objects, A and B, that are moving along the same axis.

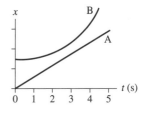

   a. At the instant $t = 1$ s, is the speed of A greater than, less than, or equal to the speed of B? Explain.

   b. Do objects A and B ever have the *same* speed? If so, at what time or times? Explain.

6. Below are six position-versus-time graphs. For each, draw the corresponding velocity-versus-time graph directly below it. A vertical line drawn through both graphs should connect the velocity v_s at time t with the position s at the *same* time t. There are no numbers, but your graphs should correctly indicate the *relative* speeds.

   a.

   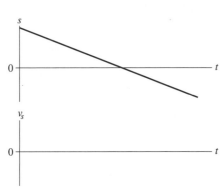

   b.

   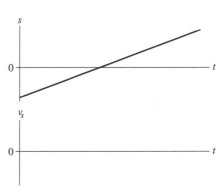

c.

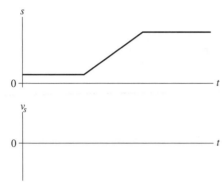

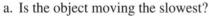

d.

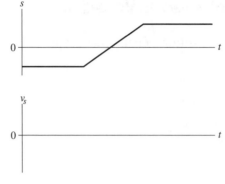

e.

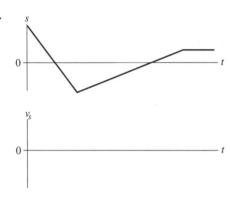

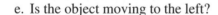

f.

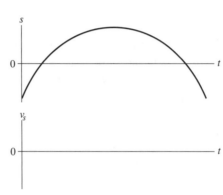

7. The figure shows a position-versus-time graph for a moving object. At which lettered point or points:

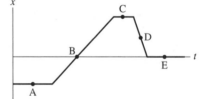

  a. Is the object moving the slowest?   _____

  b. Is the object moving the fastest?   _____

  c. Is the object at rest?   _____

  d. Does the object have a constant nonzero velocity?   _____

  e. Is the object moving to the left?   _____

8. The figure shows a position-versus-time graph for a moving object. At which lettered point or points:

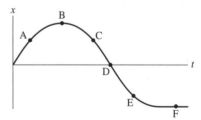

  a. Is the object moving the fastest?   _____

  b. Is the object moving to the left?   _____

  c. Is the object speeding up?   _____

  d. Is the object slowing down?   _____

  e. Is the object turning around?   _____

9. For each of the following motions, draw
   - A motion diagram,
   - A position-versus-time graph, and
   - A velocity-versus-time graph.

   a. A car starts from rest, steadily speeds up to 40 mph in 15 s, moves at a constant speed for 30 s, then comes to a halt in 5 s.

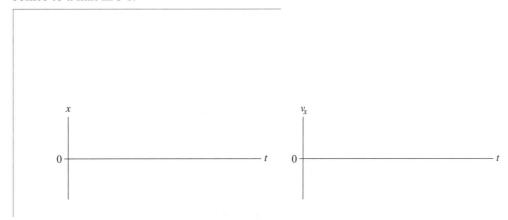

   b. A rock is dropped from a bridge and steadily speeds up as it falls. It is moving at 30 m/s when it hits the ground 3 s later. Think carefully about the signs.

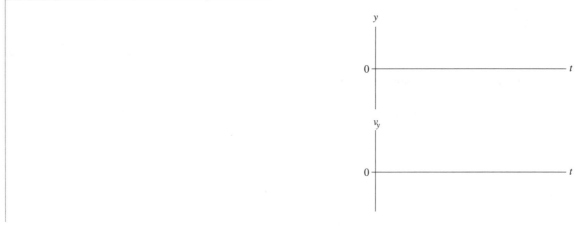

   c. A pitcher winds up and throws a baseball with a speed of 40 m/s. One-half second later the batter hits a line drive with a speed of 60 m/s. The ball is caught 1 s after it is hit. From where you are sitting, the batter is to the right of the pitcher. Draw your motion diagram and graph for the *horizontal* motion of the ball.

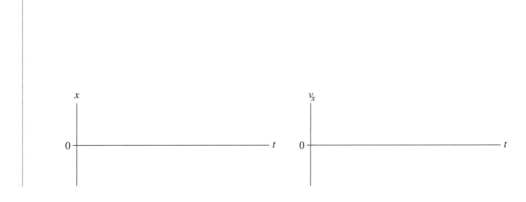

10. The figure shows six frames from the motion diagram of two moving cars, A and B.

    a. Draw both a position-versus-time graph and a velocity-versus-time graph. Show the motion of *both* cars on each graph. Label them A and B.

    b. Do the two cars ever have the same position at one instant of time?

       If so, in which frame number (or numbers)? _____

       Draw a vertical line through your graphs of part a to indicate this instant of time.

    c. Do the two cars ever have the same velocity at one instant of time?

       If so, between which two frames? _____

11. The figure shows six frames from the motion diagram of two moving cars, A and B.

    a. Draw both a position-versus-time graph and a velocity-versus-time graph. Show *both* cars on each graph. Label them A and B.

    b. Do the two cars ever have the same position at one instant of time?

       If so, in which frame number (or numbers)? _____

       Draw a vertical line through your graphs of part a to indicate this instant of time.

    c. Do the two cars ever have the same velocity at one instant of time?

       If so, between which two frames? _____

2.3 Finding Position from Velocity

12. Below are shown four velocity-versus-time graphs. For each:
    - Draw the corresponding position-versus-time graph.
    - Give a written description of the motion.

    Assume that the motion takes place along a horizontal line and that $x_0 = 0$.

a.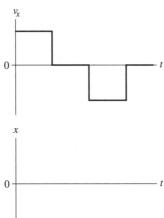

b.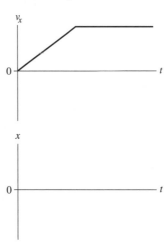

c.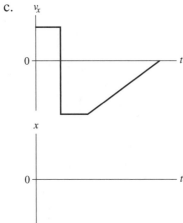

d.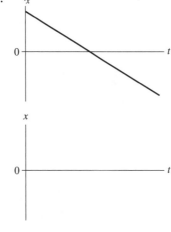

13. The figure shows the velocity-versus-time graph for a moving object whose initial position is $x_0 = 20$ m. Find the object's position graphically, using the geometry of the graph, at the following times.

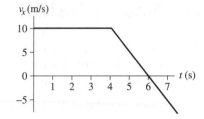

a. At $t = 3$ s.

b. At $t = 5$ s.

c. At $t = 7$ s.

d. You should have found a simple relationship between your answers to parts b and c. Can you explain this? What is the object doing?

2.4 Motion with Constant Acceleration

14. Give a specific example for each of the following situations. For each, provide:
    - A description, and
    - A motion diagram.

    a. $a_x = 0$ but $v_x \neq 0$.

    b. $v_x = 0$ but $a_x \neq 0$.

    c. $v_x < 0$ and $a_x > 0$.

15. Below are three velocity-versus-time graphs. For each:
    - Draw the corresponding acceleration-versus-time graph.
    - Draw a motion diagram below the graphs.

    a.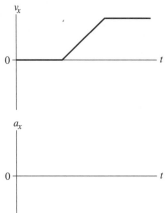

    b.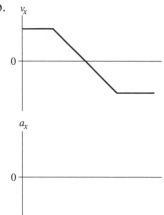

    c.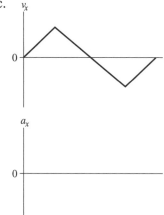

16. Below are three acceleration-versus-time graphs. For each, draw the corresponding velocity-versus-time graph. Assume that $v_{0x} = 0$.

a.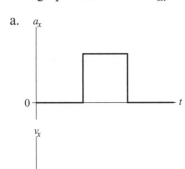

b.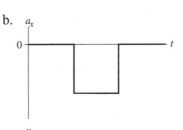

c.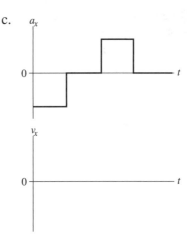

17. The figure below shows nine frames from the motion diagram of two cars. Both cars begin to accelerate, with constant acceleration, in frame 3.

a. Which car has the largest initial velocity? _____ The largest final velocity? _____

b. Which car has the largest acceleration after frame 3? How can you tell?

_____

c. Draw position, velocity, and acceleration graphs, showing the motion of both cars on each graph. (Label them A and B.) This is a total of three graphs with two curves on each.

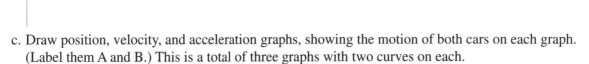

d. Do the cars ever have the same position at one instant of time? If so, in which frame? _____

e. Do the two cars ever have the same velocity at one instant of time? _____

If so, identify the *two* frames between which this velocity occurs. _____

Identify this instant on your graphs by drawing a vertical line through the graphs.

2.5 Free Fall

18. A ball is thrown straight up into the air. At each of the following instants, is the magnitude of the ball's acceleration greater than g, equal to g, less than g, or zero?

    a. Just after leaving your hand? _____

    b. At the very top (maximum height)? _____

    c. Just before hitting the ground? _____

19. A rock is *thrown* (not dropped) straight down from a bridge into the river below.

    a. Immediately *after* being released, is the magnitude of the rock's acceleration greater than g, less than g, or equal to g? Explain.

    b. Immediately before hitting the water, is the magnitude of the rock's acceleration greater than g, less than g, or equal to g? Explain.

20. Alicia throws a red ball straight up into the air, releasing it with velocity v_0. As she is throwing it, you happen to pass by in an elevator that is rising with constant velocity v_0. At the exact instant Alicia releases her ball, you reach out of the elevator's window (this is a very fancy elevator!) and *gently* release a blue ball. Both balls are the same height above the ground at the moment they are released.

    a. Describe the motion of the two balls as Alicia sees them from the ground. In what ways are the motion of the red ball and the blue ball the same or different?

b. Describe the motion of the two balls as you see them from the moving elevator. In what ways are the motion of the red ball and the blue ball the same or different?

c. Alicia sees a well-defined "top" of the motion where her red ball reaches a maximum height and then starts to fall. Call the time of maximum height t_1. As you watch from the elevator, do *you* see anything distinctive or different about the red ball's motion at time t_1? If so, what?

d. Does the red ball "stop" at time t_1 when Alice sees it at the very top of its trajectory? As part of answering this question, define what you mean by the word "stop."

2.6 Motion on an Inclined Plane

21. A ball released from rest on an inclined plane accelerates down the plane at 2 m/s^2. Complete the table below showing the ball's velocities at the times indicated. Do *not* use a calculator for this; this is a reasoning question, not a calculation problem.

Time (s)	Velocity (m/s)
0	0
1	_____
2	_____
3	_____
4	_____
5	_____

22. A bowling ball rolls along a level surface, then up a 30° slope, and finally exits onto another level surface at a much slower speed.

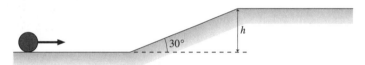

a. Draw position-, velocity-, and acceleration-versus-time graphs for the ball.

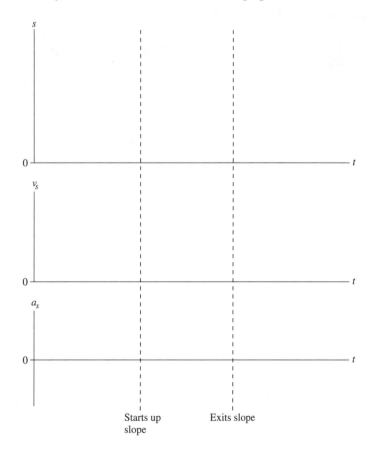

b. Suppose that the ball's initial speed is 5.0 m/s and its final speed is 1.0 m/s. Draw a pictorial representation that you would use to determine the height h of the slope. Establish a coordinate system, define all symbols, list known information, and identify desired unknowns.

Note: Don't actually solve the problem. Just draw the complete pictorial representation that you would use as a first step in solving the problem.

2.7 Instantaneous Acceleration

23. Below are two acceleration-versus-time curves. For each, draw the corresponding velocity-versus-time curve. Assume that $v_{0x} = 0$.

a.

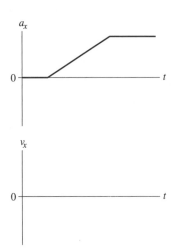

b.

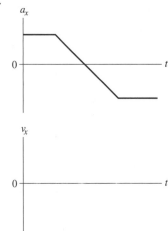

3 Vectors and Coordinate Systems

3.1 Vectors

3.2 Properties of Vectors

Exercises 1–3: Draw and label the vector sum $\vec{A} + \vec{B}$.

1.

2.

3.

4. Use a figure and the properties of vector addition to show that vector addition is associative. That is, show that

$$(\vec{A} + \vec{B}) + \vec{C} = \vec{A} + (\vec{B} + \vec{C})$$

Exercises 5–7: Draw and label the vector difference $\vec{A} - \vec{B}$.

5.

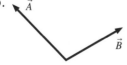

6.

7.

8. Draw and label the vector $2\vec{A}$ and the vector $\frac{1}{2}\vec{A}$.

9. Given vectors $\vec{A}$ and $\vec{B}$ below, find the vector $\vec{C} = 2\vec{A} - 3\vec{B}$.

3.3 Coordinate Systems and Vector Components

Exercises 10–12: Draw and label the *x*- and *y*-component vectors of the vector shown.

10.

11.

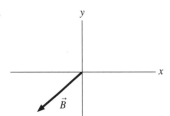

12.

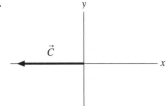

Exercises 13–15: Determine the numerical values of the *x*- and *y*-components of each vector.

13.

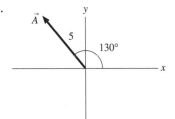

$A_x =$ _____

$A_y =$ _____

14.

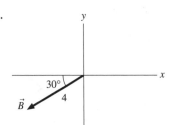

$B_x =$ _____

$B_y =$ _____

15.

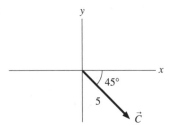

$C_x =$ _____

$C_y =$ _____

Exercises 16–18: Draw and label the vector with these components. Then determine the magnitude of the vector.

16. $A_x = 3, A_y = -2$

17. $B_x = -2, B_y = 2$

18. $C_x = 0, C_y = -2$

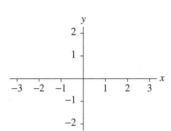

$A =$ _____

$B =$ _____

$C =$ _____

3.4 Vector Algebra

Exercises 19–21: Draw and label the vectors on the axes.

19. $\vec{A} = -\hat{i} + 2\hat{j}$

20. $\vec{B} = -2\hat{j}$

21. $\vec{C} = 3\hat{i} - 2\hat{j}$

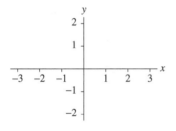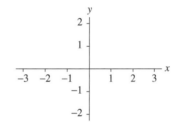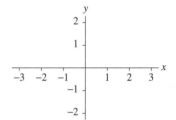

Exercises 22–24: Write the vector in component form (e.g., $3\hat{i} + 2\hat{j}$).

22.

23.

24.

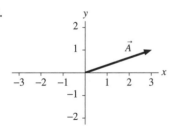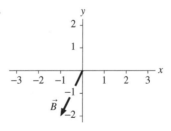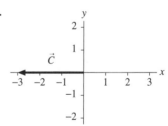

$\vec{A} =$ _____

$\vec{B} =$ _____

$\vec{C} =$ _____

25. What is the vector sum $\vec{D} = \vec{A} + \vec{B} + \vec{C}$ of the three vectors defined in Exercises 22–24? Write your answer in *component* form.

Exercises 26–28: For each vector:
- Draw the vector on the axes provided.
- Draw and label an angle θ to describe the direction of the vector.
- Find the magnitude and the angle of the vector.

26. $\vec{A} = 2\hat{i} + 2\hat{j}$

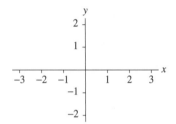

$A = $ _____

$\theta = $ _____

27. $\vec{B} = -2\hat{i} + 2\hat{j}$

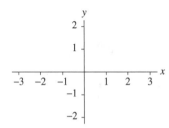

$B = $ _____

$\theta = $ _____

28. $\vec{C} = 3\hat{i} + \hat{j}$

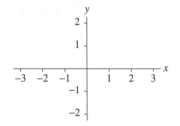

$C = $ _____

$\theta = $ _____

Exercises 29–31: Define vector $\vec{A} = (5, 30°$ above the horizontal). Determine the components A_x and A_y in the three coordinate systems shown below. Show your work below the figure.

29.

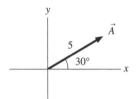

$A_x = $ _____

$A_y = $ _____

30.

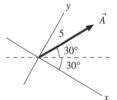

$A_x = $ _____

$A_y = $ _____

31.

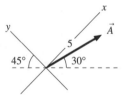

$A_x = $ _____

$A_y = $ _____

4 Kinematics in Two Dimensions

4.1 Acceleration

Exercises 1–2: The figures below show an object's position in three successive frames of film. The object is moving in the direction $0 \rightarrow 1 \rightarrow 2$. For each diagram:
- Draw and label the initial and final velocity vectors $\vec{v}_0$ and $\vec{v}_1$. Use **black**.
- Use the steps of Figures 4.2 and 4.3 to find the change in velocity $\Delta \vec{v}$.
- Draw and label $\vec{a}$ at the proper location on the motion diagram. Use **red**.
- Determine whether the object is speeding up, slowing down, or moving at a constant speed. Write your answer beside the diagram.

1.

2.

3. The figure shows a ramp and a ball that rolls along the ramp. Draw vector arrows on the figure to show the ball's acceleration at each of the lettered points A to E (or write $\vec{a} = \vec{0}$, if appropriate).

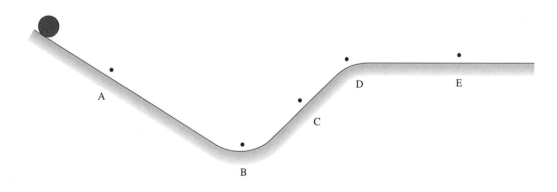

4. Complete the motion diagram for this trajectory, showing velocity and acceleration vectors.

Exercises 5–6: Draw a complete motion diagram for each of the following.
- Draw and label the velocity vectors $\vec{v}$. Use **black**.
- Draw and label the acceleration vectors $\vec{a}$. Use **red**.

5. A cannon ball is fired from a Civil War cannon up onto a high cliff. Show the cannon ball's motion from the instant it leaves the cannon until a microsecond before it hits the ground.

6. A plane flying north at 300 mph turns slowly to the west without changing speed, then continues to fly west. Draw the motion diagram from a viewpoint above the plane.

4.2 Kinematics in Two Dimensions

7. A particle moving in the *xy*-plane has the *x*-versus-*t* graph and the *y*-versus-*t* graphs shown below. Use the grid to draw a *y*-versus-*x* graph of the trajectory.

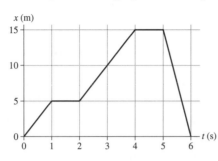
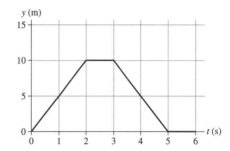

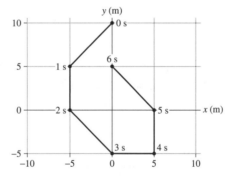

8. The trajectory of a particle is shown below. The particle's position is indicated with dots at 1-second intervals. The particle moves between each pair of dots at constant speed. Draw *x*-versus-*t* and *y*-versus-*t* graphs for the particle.

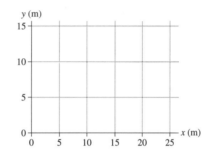

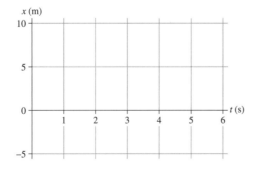
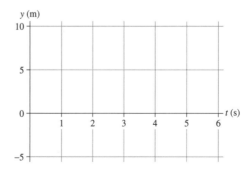

4.3 Projectile Motion

9. The figure shows a ball that rolls down a quarter-circle ramp, then off a cliff. Sketch the ball's trajectory from the instant it is released until it hits the ground.

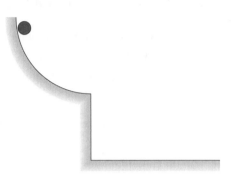

10. a. A cart that is rolling at constant velocity fires a ball straight up. When the ball comes back down, will it land in front of the launching tube, behind the launching tube, or directly in the tube? Explain.

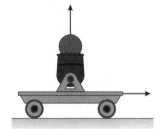

   b. Will your answer change if the cart is accelerating in the forward direction? If so, how?

11. Rank in order, from shortest to longest, the amount of time it takes each of these projectiles to hit the ground. Ignore air resistance. (Some may be simultaneous.)

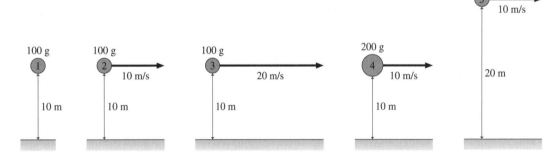

Order:

Explanation:

4.4 Relative Motion

12. Anita is running to the right at 5 m/s. Balls 1 and 2 are thrown toward her at 10 m/s by friends standing on the ground. According to Anita, which ball is moving faster? Or are both speeds the same? Explain.

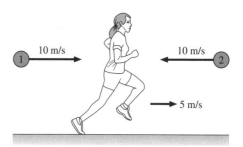

13. Anita is running to the right at 5 m/s. Balls 1 and 2 are thrown toward her by friends standing on the ground. According to Anita, both balls are approaching her at 10 m/s. Which ball was thrown at a faster speed? Or were they thrown with the same speed? Explain.

14. Ryan, Samantha, and Tomas are driving their convertibles. At the same instant, they each see a jet plane with an instantaneous velocity of 200 m/s and an acceleration of 5 m/s².

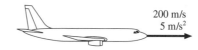

 a. Rank in order, from largest to smallest, the jet's *speed* v_R, v_S, and v_T according to Ryan, Samantha, and Tomas. Explain.

 b. Rank in order, from largest to smallest, the jet's *acceleration* a_R, a_S, and a_T according to Ryan, Samantha, and Tomas. Explain.

15. An electromagnet on the ceiling of an airplane holds a steel ball. When a button is pushed, the magnet releases the ball. The experiment is first done while the plane is parked on the ground, and the point where the ball hits the floor is marked with an X. Then the experiment is repeated while the plane is flying level at a steady 500 mph. Does the ball land slightly in front of the X (toward the nose of the plane), on the X, or slightly behind the X (toward the tail of the plane)? Explain.

16. Zack is driving past his house. He wants to toss his physics book out the window and have it land in his driveway. If he lets go of the book exactly as he passes the end of the driveway, should he direct his throw outward and toward the front of the car (throw 1), straight outward (throw 2), or outward and toward the back of the car (throw 3)? Explain. (Ignore air resistance.)

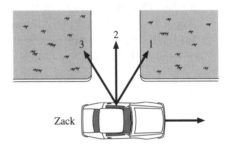

17. Yvette and Zack are driving down the freeway side by side with their windows rolled down. Zack wants to toss his physics book out the window and have it land in Yvette's front seat. Should he direct his throw outward and toward the front of the car (throw 1), straight outward (throw 2), or outward and toward the back of the car (throw 3)? Explain. (Ignore air resistance.)

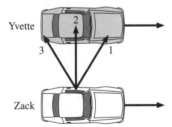

4.5 Uniform Circular Motion

4.6 Velocity and Acceleration in Uniform Circular Motion

18. a. The crankshaft in your car rotates at 3000 rpm. What is the frequency in revolutions per second?

    b. A record turntable rotates at 33.3 rpm. What is the period in seconds?

19. The figure shows three points on a steadily rotating wheel.

    a. Draw the velocity vectors at each of the three points.

    b. Rank in order, from largest to smallest, the angular velocities ω_1, ω_2, and ω_3 of these points.

    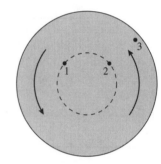

       Order:

       Explanation:

    c. Rank in order, from largest to smallest, the speeds v_1, v_2, and v_3 of these points.

       Order:

       Explanation:

20. Below are two angular position-versus-time graphs. For each, draw the corresponding angular velocity-versus-time graph directly below it.

a.

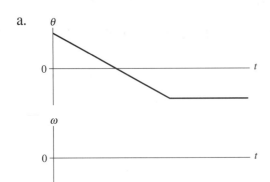

b.

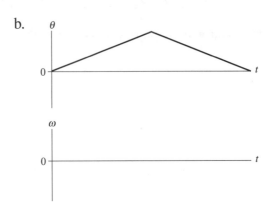

21. Below are two angular velocity-versus-time graphs. For each, draw the corresponding angular position-versus-time graph directly below it. Assume $\theta_0 = 0$ rad.

a.

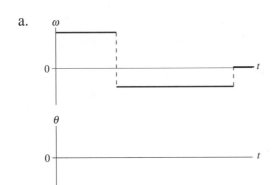

b.

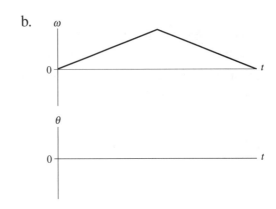

22. A particle in circular motion rotates clockwise at 4 rad/s for 2 s, then counterclockwise at 2 rad/s for 4 s. The time required to change direction is negligible. Graph the angular velocity and the angular position, assuming $\theta_0 = 0$ rad.

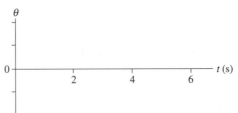

23. A particle rotates in a circle with $a_r = 8$ m/s^2. What is a_r if

   a. The radius is doubled without changing the angular velocity?   _____

   b. The radius is doubled without changing the particle's speed?   _____

   c. The angular velocity is doubled without changing the particle's radius?   _____

4.7 Nonuniform Circular Motion and Angular Acceleration

24. The following figures show a rotating wheel. Determine the signs (+ or −) of ω and α.

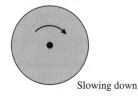 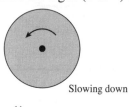 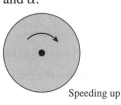

Speeding up   Slowing down   Slowing down   Speeding up

ω .... ω .... ω .... ω ....

α .... α .... α .... α ....

25. The figures below show the radial acceleration vector $\vec{a}_r$ at four successive points on the trajectory of a particle moving in a counterclockwise circle.

   a. For each, draw the tangential acceleration vector $\vec{a}_t$ at points 2 and 3 or, if appropriate, write $\vec{a}_t = \vec{0}$.
   b. Determine if the particle's angular acceleration α is positive (+), negative (−), or zero (0).

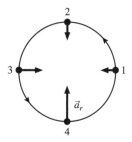 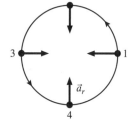 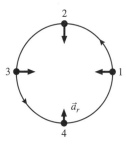

$\alpha =$ _____ $\alpha =$ _____ $\alpha =$ _____

26. A pendulum swings from its end point on the left (point 1) to its end point on the right (point 5). At each of the labeled points:

   a. Use a **black** pen or pencil to draw and label the vectors $\vec{a}_r$ and $\vec{a}_t$ at each point. Make sure the length indicates the relative size of the vector.

   b. Use a **red** pen or pencil to draw and label the total acceleration vector $\vec{a}$.

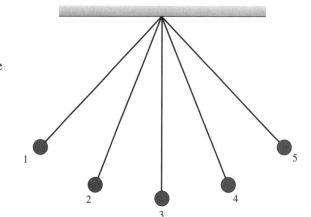

27. The figure shows the θ-versus-t graph for a particle moving in a circle. The curves are all sections of parabolas.

    a. Draw the corresponding ω-versus-t and α-versus-t graphs. Notice that the horizontal tick marks are equally spaced.

    b. Write a description of the particle's motion.

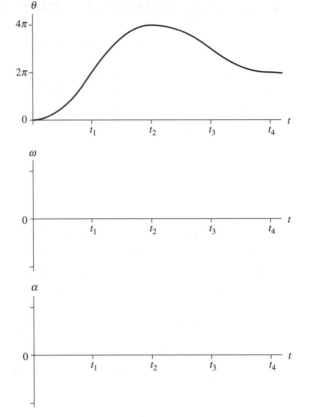

28. A wheel rolls to the left along a horizontal surface, up a ramp, then continues along the upper horizontal surface. Draw graphs for the wheel's angular velocity ω and angular acceleration α as functions of time.

5 Force and Motion

5.1 Force

1. Two or more forces are shown on the objects below. Draw and label the net force $\vec{F}_{net}$.

2. Two or more forces are shown on the objects below. Draw and label the net force $\vec{F}_{net}$.

5.2 A Short Catalog of Forces

5.3 Identifying Forces

Exercises 3–8: Follow the six-step procedure of Tactics Box 5.2 to identify and name all the forces acting on the object.

3. An elevator suspended by a cable is descending at constant velocity.

4. A car on a *very* slippery icy road is sliding headfirst into a snowbank, where it gently comes to rest with no one injured. (Question: What does "*very* slippery" imply?)

5. A compressed spring is pushing a block across a rough horizontal table.

6. A brick is falling from the roof of a three-story building.

7. Blocks A and B are connected by a string passing over a pulley. Block B is falling and dragging block A across a frictionless table. Analyze block A.

8. A rocket is launched at a 30° angle. Air resistance is not negligible.

5.4 What Do Forces Do? A Virtual Experiment

9. The figure shows an acceleration-versus-force graph for an object of mass m. Data have been plotted as individual points, and a line has been drawn through the points.

   Draw and label, directly on the figure, the acceleration-versus-force graphs for objects of mass

   a. $2m$ b. $0.5m$

   Use triangles ▲ to show four points for the object of mass $2m$, then draw a line through the points. Use squares ■ for the object of mass $0.5m$.

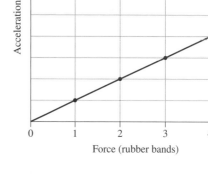

10. A constant force applied to object A causes A to accelerate at 5 m/s². The same force applied to object B causes an acceleration of 3 m/s². Applied to object C, it causes an acceleration of 8 m/s².

   a. Which object has the largest mass? _____

   b. Which object has the smallest mass? _____

   c. What is the ratio of mass A to mass B? $(m_A/m_B) =$ _____

11. A constant force applied to an object causes the object to accelerate at 10 m/s². What will the acceleration of this object be if

   a. The force is doubled? _____   b. The mass is doubled? _____

   c. The force is doubled *and* the mass is doubled? _____

   d. The force is doubled *and* the mass is halved? _____

12. A constant force applied to an object causes the object to accelerate at 8 m/s². What will the acceleration of this object be if

   a. The force is halved? _____   b. The mass is halved? _____

   c. The force is halved *and* the mass is halved? _____

   d. The force is halved *and* the mass is doubled? _____

5.5 Newton's Second Law

13. Forces are shown on two objects. For each:
   a. Draw and label the net force vector. Do this right on the figure.
   b. Below the figure, draw and label the object's acceleration vector.

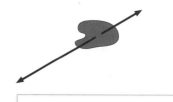

14. Forces are shown on two objects. For each:
    a. Draw and label the net force vector. Do this right on the figure.
    b. Below the figure, draw and label the object's acceleration vector.

15. In the figures below, one force is missing. Use the given direction of acceleration to determine the missing force and draw it on the object. Do all work directly on the figure.

16. Below are two motion diagrams for a particle. Draw and label the net force vector at point 2.

17. Below are two motion diagrams for a particle. Draw and label the net force vector at point 2.

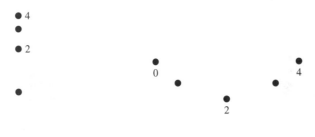

5.6 Newton's First Law

18. If an object is at rest, can you conclude that there are no forces acting on it? Explain.

19. If a force is exerted on an object, is it possible for that object to be moving with constant velocity? Explain.

20. A hollow tube forms three-quarters of a circle. It is lying flat on a table. A ball is shot through the tube at high speed. As the ball emerges from the other end, does it follow path A, path B, or path C? Explain your reasoning.

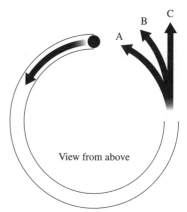

View from above

21. Which, if either, of the objects shown below is in equilibrium? Explain your reasoning.

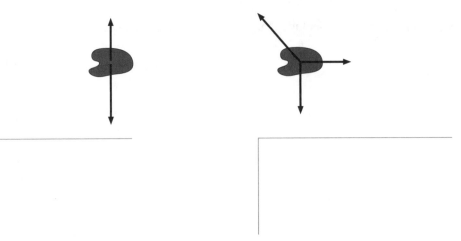

22. Two forces are shown on the objects below. Add a third force $\vec{F}_3$ that will cause the object to be in equilibrium.

23. Are the following inertial reference frames? Answer Yes or No.

    a. A car driving at steady speed on a straight and level road.

    b. A car driving at steady speed up a 10° incline.

    c. A car speeding up after leaving a stop sign.

    d. A car driving at steady speed around a curve.

    e. A hot air balloon rising straight up at steady speed.

    f. A skydiver just after leaping out of a plane.

    g. The space shuttle orbiting the earth.

5.7 Free-Body Diagrams

Exercises 24–29:
- Draw a picture and identify the forces, then
- Draw a complete free-body diagram for the object, following each of the steps given in Tactics Box 5.3. Be sure to think carefully about the direction of $\vec{F}_{net}$.

Note: Draw individual force vectors with a **black** or **blue** pencil or pen. Draw the *net* force vector $\vec{F}_{net}$ with a **red** pencil or pen.

24. A heavy crate is being lowered straight down at a constant speed by a steel cable.

25. A boy is pushing a box across the floor at a steadily increasing speed. Let the box be "the system" for analysis.

26. A bicycle is speeding up down a hill. Friction is negligible, but air resistance is not.

27. You've slammed on your car brakes while going down a hill. You're skidding to a halt.

28. You are going to toss a rock *straight up* into the air by placing it on the palm of your hand (you're not gripping it), then pushing your hand up very rapidly. You may want to toss an object into the air this way to help you think about the situation. The rock is "the system" of interest.

    a. As you hold the rock at rest on your palm, before moving your hand.

    b. As your hand is moving up but before the rock leaves your hand.

    c. One-tenth of a second after the rock leaves your hand.

    d. After the rock has reached its highest point and is now falling straight down.

29. Block B has just been released and is beginning to fall. The table has friction. Analyze block A.

6 Dynamics I: Motion Along a Line

6.1 Equilibrium

1. The vectors below show five forces that can be applied individually or in combinations to an object. Which forces or combinations of forces will cause the object to be in equilibrium?

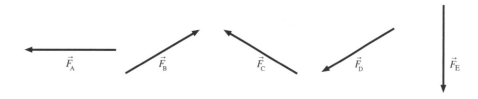

2. The free-body diagrams show a force or forces acting on an object. Draw and label one more force (one that is appropriate to the situation) that will cause the object to be in equilibrium.

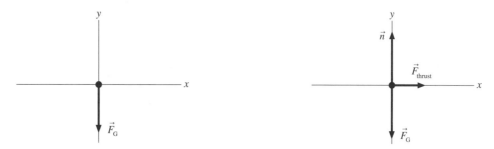

3. If you know all of the forces acting on a moving object, can you tell in which direction the object is moving? If the answer is Yes, explain how. If the answer is No, give an example.

6.2 Using Newton's Second Law

4. a. An elevator travels *upward* at a constant speed. The elevator hangs by a single cable. Friction and air resistance are negligible. Is the tension in the cable greater than, less than, or equal to the weight of the elevator? Explain. Your explanation should include both a free-body diagram and reference to appropriate physical principles.

b. The elevator travels *downward* and is slowing down. Is the tension in the cable greater than, less than, or equal to the weight of the elevator? Explain.

Exercises 5–6: The figures show free-body diagrams for an object of mass *m*. Write the *x*- and *y*-components of Newton's second law. Write your equations in terms of the *magnitudes* of the forces $F_1, F_2, \ldots$ and any *angles* defined in the diagram. One equation is shown to illustrate the procedure.

5.

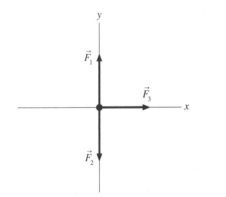

$ma_x = $ _____

$ma_y = F_1 - F_2$

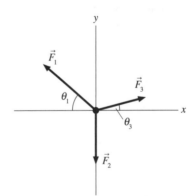

$ma_x = $ _____

$ma_y = $ _____

6.

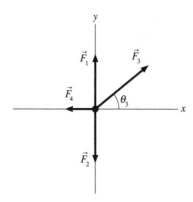

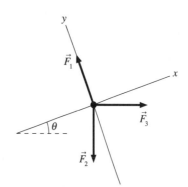

$$ma_x = F_3 \cos\theta_3 - F_4$$

$$ma_y = \underline{\hspace{5cm}}$$

$$ma_x = \underline{\hspace{5cm}}$$

$$ma_y = \underline{\hspace{5cm}}$$

Exercises 7–9: Two or more forces, shown on a free-body diagram, are exerted on a 2 kg object. The units of the grid are newtons. For each:

- Draw a vector arrow *on the grid,* starting at the origin, to show the net force $\vec{F}_{net}$.
- In the space to the right, determine the numerical values of the components a_x and a_y.

7.

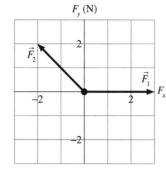

$a_x = \underline{\hspace{5cm}}$

$a_y = \underline{\hspace{5cm}}$

8.

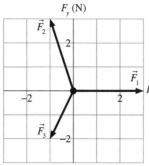

$a_x = \underline{\hspace{5cm}}$

$a_y = \underline{\hspace{5cm}}$

9.

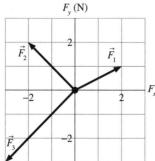

$a_x = \underline{\hspace{5cm}}$

$a_y = \underline{\hspace{5cm}}$

Exercises 10–12: Three forces $\vec{F}_1$, $\vec{F}_2$, and $\vec{F}_3$ cause a 1 kg object to accelerate with the acceleration given. Two of the forces are shown on the free-body diagrams below, but the third is missing. For each, draw and label *on the grid* the missing third force vector.

10. $\vec{a} = 2\hat{\imath}\ \text{m/s}^2$

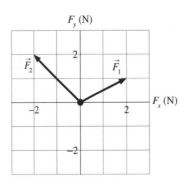

11. $\vec{a} = -3\hat{\jmath}\ \text{m/s}^2$

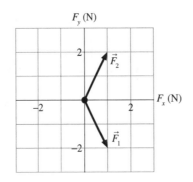

12. The object moves with constant velocity.

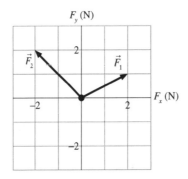

13. Three arrows are shot horizontally. They have left the bow and are traveling parallel to the ground. Air resistance is negligible. Rank in order, from largest to smallest, the magnitudes of the *horizontal* forces F_1, F_2, and F_3 acting on the arrows. Some may be equal. Give your answer in the form A > B = C > D.

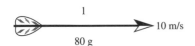 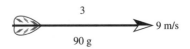

Order:

Explanation:

6.3 Mass, Weight, and Gravity

14. An astronaut takes his bathroom scales to the moon and then stands on them. Is the reading of the scales his weight? Explain.

15. Suppose you attempt to pour out 100 g of salt, using a pan balance for measurement, while in an elevator that is accelerating upward. Will the quantity of salt be too much, too little, or the correct amount? Explain.

16. An astronaut orbiting the earth is handed two balls that are identical in outward appearance. However, one is hollow while the other is filled with lead. How might the astronaut determine which is which? Cutting them open is not allowed.

© 2008 by Pearson Education, Inc., publishing as Pearson Addison-Wesley.

17. The terms "vertical" and "horizontal" are frequently used in physics. Give operational definitions for these two terms. An operational definition defines a term by how it is measured or determined. Your definition should apply equally well in a laboratory or on a steep mountainside.

18. Suppose you stand on a spring scale in six identical elevators. Each elevator moves as shown below. Let the reading of the scale in elevator n be S_n. Rank in order, from largest to smallest, the six scale readings S_1 to S_6. Some may be equal. Give your answer in the form $A > B = C > D$.

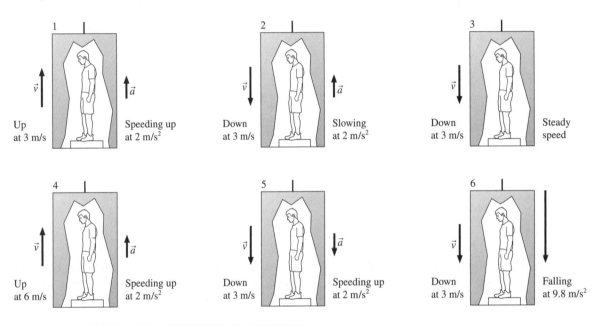

Order:

Explanation:

6.4 Friction

19. A block pushed along the floor with velocity $\vec{v}_0$ slides a distance d after the pushing force is removed.

    a. If the mass of the block is doubled but the initial velocity is not changed, what is the distance the block slides before stopping? Explain.

    b. If the initial velocity of the block is doubled to $2\vec{v}_0$ but the mass is not changed, what is the distance the block slides before stopping? Explain.

20. Suppose you press a book against the wall with your hand. The book is not moving.

    a. Identify the forces on the book and draw a free-body diagram.

    b. Now suppose you decrease your push, but not enough for the book to slip. What happens to each of the following forces? Do they increase in magnitude, decrease, or not change?

    $\vec{F}_{push}$ _____

    $\vec{F}_{G}$ _____

    $\vec{n}$ _____

    $\vec{f}_{s}$ _____

    $\vec{f}_{s\,max}$ _____

6.5 Drag

21. Consider a box in the back of a pickup truck.

    a. If the truck accelerates slowly, the box moves with the truck without slipping. What force or forces act on the box to accelerate it? In what direction do those forces point?

    b. Draw a free-body diagram of the box.

    c. What happens to the box if the truck accelerates too rapidly? Explain why this happens, basing your explanation on physical models and the principles described in this chapter.

22. Three objects move through the air as shown. Rank in order, from largest to smallest, the three drag forces D_1, D_2, and D_3. Some may be equal. Give your answer in the form A > B = C > D.

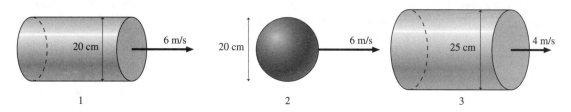

1   2   3

Order:

Explanation:

23. Five balls move through the air as shown. All five have the same size and shape. Rank in order, from largest to smallest, the magnitude of their accelerations a_1 to a_5. Some may be equal. Give your answer in the form A > B = C > D.

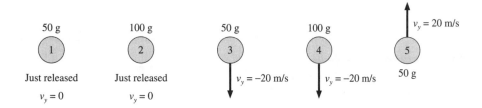

Order:

Explanation:

24. A 1 kg wood ball and a 10 kg lead ball have identical shapes and sizes. They are dropped simultaneously from a tall tower.

    a. To begin, assume that air resistance is negligible. As the balls fall, are the forces on them equal in magnitude or different? If different, which has the larger force? Explain.

    b. Are their accelerations equal or different? If different, which has the larger acceleration? Explain.

    c. Which ball hits the ground first? Or do they hit simultaneously? Explain.

    d. If air resistance is present, each ball will experience the *same* drag force because both have the same shape. Draw free-body diagrams for the two balls as they fall in the presence of air resistance. Make sure that your vectors all have the correct *relative* lengths.

    e. When air resistance is included, are the accelerations of the balls equal or different? If not, which has the larger acceleration? Explain, using your free-body diagrams and Newton's laws.

    f. Which ball now hits the ground first? Or do they hit simultaneously? Explain.

7 Newton's Third Law

7.1 Interacting Objects

7.2 Analyzing Interacting Objects

Exercises 1–7: Follow steps 1–3 of Tactics Box 7.1 to draw interaction diagrams describing the following situations. Your diagrams should be similar to Figures 7.6 and 7.10.

1. A bat hits a ball.

2. A massless string pulls a box across the floor. Friction is not negligible.

3. A boy pulls a wagon by a rope attached to the front of the wagon. The rope is not massless, and rolling friction is not negligible.

© 2008 by Pearson Education, Inc., publishing as Pearson Addison-Wesley.

4. A skateboarder is pushing on the ground to speed up. Treat the person and the skateboard as separate objects.

5. The bottom block is pulled by a massless string. Friction is not negligible. Treat the two blocks as separate objects.

6. A crate in the back of a truck does not slip as the truck accelerates forward. Treat the crate and the truck as separate objects.

7. The bottom block is pulled by a massless string. Friction is not negligible. Treat the pulley as a separate object.

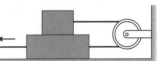

7.3 Newton's Third Law

8. Block A is pushed across a horizontal surface at a *constant* speed by a hand that exerts force $\vec{F}_{\text{H on A}}$. The surface has friction.

Hand

A

a. Draw two free-body diagrams, one for the hand and the other for the block. On these diagrams:

- Show only the *horizontal* forces, such as was done in Figure 7.14 of the text.
- Label force vectors, using the form $\vec{F}_{\text{C on D}}$.
- Connect action/reaction pairs with dotted lines.
- On the hand diagram show only $\vec{F}_{\text{A on H}}$. Don't include $\vec{F}_{\text{body on H}}$.
- Make sure vector lengths correctly portray the relative magnitudes of the forces.

b. Rank in order, from largest to smallest, the magnitudes of *all* of the horizontal forces you showed in part a. For example, if $F_{\text{C on D}}$ is the largest of three forces while $F_{\text{D on C}}$ and $F_{\text{D on E}}$ are smaller but equal, you can record this as $F_{\text{C on D}} > F_{\text{D on C}} = F_{\text{D on E}}$.

Order:

Explanation:

c. Repeat both part a and part b for the case that the block is *speeding up*.

9. A second block B is placed in front of Block A of question 8. B is more massive than A: $m_B > m_A$. The blocks are speeding up.

a. Consider a *frictionless* surface. Draw *separate* free-body diagrams for A, B, and the hand. Show only the horizontal forces. Label forces in the form $\vec{F}_{C \text{ on } D}$. Use dashed lines to connect action/reaction pairs.

b. By applying the second law to each block and the third law to each action/reaction pair, rank in order *all* of the horizontal forces, from largest to smallest.

Order:

Explanation:

c. Repeat parts a and b if the surface has friction. Assume that A and B have the same coefficient of kinetic friction.

10. Blocks A and B are held on the palm of your outstretched hand as you lift them straight up at *constant speed*. Assume $m_B > m_A$ and that $m_{hand} = 0$.

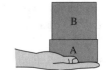

   a. Draw *separate* free-body diagrams for A, B, and your hand.
   - Show *all* vertical forces, including the gravitational forces on the blocks.
   - Make sure vector lengths indicate the relative sizes of the forces.
   - Label forces in the form $\vec{F}_{C \text{ on } D}$
   - Connect action/reaction pairs with dashed lines.

   b. Rank in order, from largest to smallest, all of the vertical forces. Explain your reasoning.

11. A mosquito collides head-on with a car traveling 60 mph.

   a. How do you think the size of the force that the car exerts on the mosquito compares to the size of the force that the mosquito exerts on the car?

   b. Draw *separate* free-body diagrams of the car and the mosquito at the moment of collision, showing only the horizontal forces. Label forces in the form $\vec{F}_{C \text{ on } D}$. Connect action/reaction pairs with dotted lines.

Exercises 12–16: Write the acceleration constraint in terms of *components*. For example, write $(a_1)_x = (a_2)_y$, if that is the appropriate answer, rather than $\vec{a}_1 = \vec{a}_2$.

12.

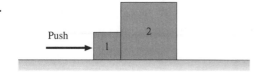

Constraint: _____

13.

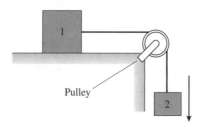

Constraint: _____

14.

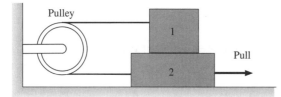

Constraint: _____

15.

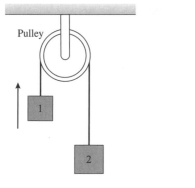

Constraint: _____

16.

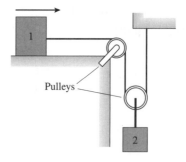

Constraint: _____

7.4 Ropes and Pulleys

Exercises 17–22: Determine the reading of the spring scale.

- All the masses are at rest.
- The strings and pulleys are massless, and the pulleys are frictionless.
- The spring scale reads in kg.

17.

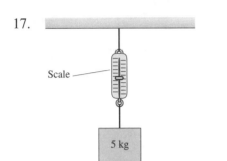

Scale = _____

18.

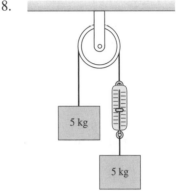

Scale = _____

19.

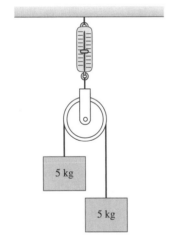

Scale = _____

20.

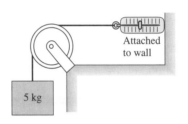

Scale = _____

21.

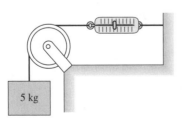

Scale = _____

22.

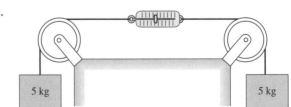

Scale = _____

7.5 Examples of Interacting-Objects Problems

23. Blocks A and B, with $m_B > m_A$, are connected by a string. A hand pushing on the back of A accelerates them along a frictionless surface. The string (S) is massless.

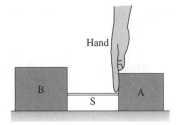

a. Draw separate free-body diagrams for A, S, and B, showing only horizontal forces. Be sure vector lengths indicate the relative size of the force. Connect any action/reaction pairs with dotted lines.

b. Rank in order, from largest to smallest, all of the horizontal forces. Explain.

c. Repeat parts a and b if the string has mass.

d. You might expect to find $F_{S\ on\ B} > F_{H\ on\ A}$ because $m_B > m_A$. Did you? Explain why $F_{S\ on\ B} > F_{H\ on\ A}$ is or is not a correct statement.

24. Blocks A and B are connected by a massless string over a massless, frictionless pulley. The blocks have just this instant been released from rest.

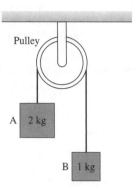

a. Will the blocks accelerate? If so, in which directions?

b. Draw a separate free-body diagram for each block. Be sure vector lengths indicate the relative size of the force. Connect any action/reaction pairs or "as if" pairs with dashed lines.

c. Rank in order, from largest to smallest, all of the vertical forces. Explain.

d. Compare the magnitude of the *net* force on A with the *net* force on B. Are they equal, or is one larger than the other? Explain.

e. Consider the block that falls. Is the magnitude of its acceleration less than, greater than, or equal to *g*? Explain.

25. In case a, block A is accelerated across a frictionless table by a hanging 10 N weight (1.02 kg). In case b, the same block is accelerated by a steady 10 N tension in the string.

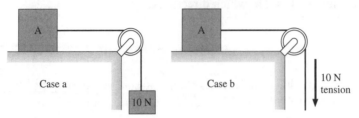

Is block A's acceleration in case b greater than, less than, or equal to its acceleration in case a? Explain.

Exercises 26–27: Draw separate free-body diagrams for blocks A and B. Connect any action/ reaction pairs (or forces that act *as if* they are action/reaction pairs) together with dashed lines.

26.

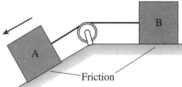

27.

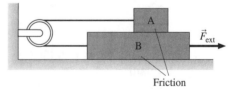

8 Dynamics II: Motion in a Plane

8.1 Dynamics in Two Dimensions

1. An ice hockey puck is pushed across frictionless ice in the direction shown. The puck receives a sharp, very short-duration kick toward the right as it crosses line 2. It receives a second kick, of equal strength and duration but toward the left, as it crosses line 3. Sketch the puck's trajectory from line 1 until it crosses line 4.

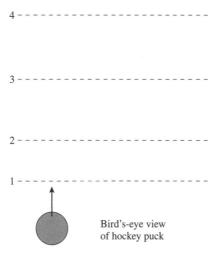

2. A rocket motor is taped to an ice hockey puck, oriented so that the thrust is to the left. The puck is given a push across frictionless ice in the direction shown. The rocket will be turned on by remote control as the puck crosses line 2, then turned off as it crosses line 3. Sketch the puck's trajectory from line 1 until it crosses line 4.

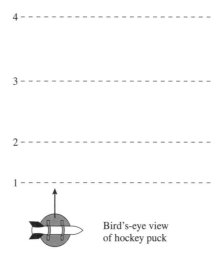

3. An ice hockey puck is sliding from west to east across frictionless ice. When the puck reaches the point marked by the dot, you're going to give it *one* sharp blow with a hammer. After hitting it, you want the puck to move from north to south at a speed similar to its initial west-to-east speed. Draw a force vector with its tail on the dot to show the direction in which you will aim your hammer blow.

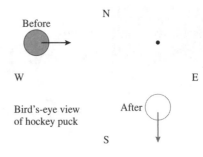

4. Tarzan swings through the jungle by hanging from a vine.
   a. Draw a motion diagram of Tarzan, as you learned in Chapter 1. Use it to find the direction of Tarzan's acceleration vector $\vec{a}$:
      i.   Immediately after stepping off the branch, and
      ii. At the lowest point in his swing.

   b. At the lowest point in the swing, is the tension T in the vine greater than, less than, or equal to Tarzan's weight? Explain, basing your explanation on Newton's laws.

8.2 Velocity and Acceleration in Uniform Circular Motion

8.3 Dynamics of Uniform Circular Motion

5. The figure shows a *top view* of a plastic tube that is fixed on a horizontal table top. A marble is shot into the tube at A. On the figure, sketch the marble's trajectory after it leaves the tube at B.

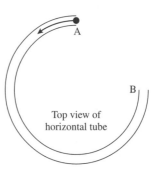

Top view of horizontal tube

6. A ball swings in a *vertical* circle on a string. During one revolution, a very sharp knife is used to cut the string at the instant when the ball is at its lowest point. Sketch the subsequent trajectory of the ball until it hits the ground.

Knife

7. The figures are a bird's-eye view of particles on a string moving in horizontal circles on a table top. All are moving at the same speed. Rank in order, from largest to smallest, the string tensions T_1 to T_4.

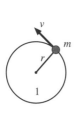
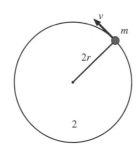
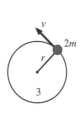
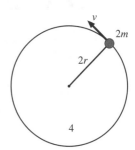

Order:

Explanation:

8. A ball on a string moves in a vertical circle. When the ball is at its lowest point, is the tension in the string greater than, less than, or equal to the ball's weight? Explain. (You may want to include a free-body diagram as part of your explanation.)

9. A marble rolls around the inside of a cone. Draw a free-body diagram of the marble when it is on the left side of the cone and a free-body diagram of the marble when it is on the right side of the cone.

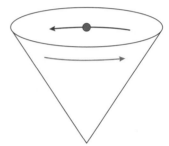

On left side   On right side

8.4 Circular Orbits

10. The earth has seasons because the axis of the earth's rotation is tilted 23° away from a line perpendicular to the plane of the earth's orbit. You can see this in the figure, which shows an edge view of the earth's orbit around the sun. For both positions of the earth, draw a force vector to show the net force acting on the earth or, if appropriate, write $\vec{F} = \vec{0}$.

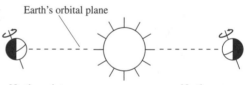

Earth's orbital plane

Norther winter   Norther summer
Southern summer   Southern winter

11. A small projectile is launched parallel to the ground at height $h = 1$ m with sufficient speed to orbit a completely smooth, airless planet. A bug rides in a small hole inside the projectile. Is the bug weightless? Explain.

8.5 Fictitious Forces

8.6 Why Does the Water Stay in the Bucket?

12. A stunt plane does a series of vertical loop-the-loops. At what point in the circle does the pilot feel the heaviest? Explain. Include a free-body diagram with your explanation.

13. You can swing a ball on a string in a *vertical* circle if you swing it fast enough.

    a. Draw two free-body diagrams of the ball at the top of the circle. On the left, show the ball when it is going around the circle very fast. On the right, show the ball as it goes around the circle more slowly.

    | Very fast | Slower |

    b. As you continue slowing the swing, there comes a frequency at which the string goes slack and the ball doesn't make it to the top of the circle. What condition must be satisfied for the ball to be able to complete the full circle?

    c. Suppose the ball has the smallest possible frequency that allows it to go all the way around the circle. What is the tension in the string when the ball is at the highest point? Explain.

14. It's been proposed that future space stations create "artificial gravity" by rotating around an axis.

   a. How would this work? Explain.

   b. Would the artificial gravity be equally effective throughout the space station? If not, where in the space station would the residents want to live and work?

8.7 Nonuniform Circular Motion

15. For each, figure determine the signs (+ or −) of ω and α.

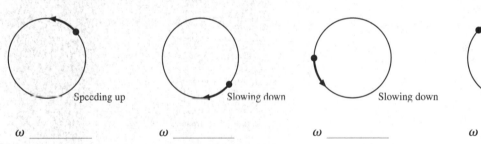

            Speeding up   Slowing down   Slowing down   Speeding up

ω _____ ω _____ ω _____ ω _____

α _____ α _____ α _____ α _____

16. The figures below show the radial acceleration vector $\vec{a}_r$ at four sequential points on the trajectory of a particle moving in a counterclockwise circle.

   a. For each, draw the tangential acceleration vector α at points 2 and 3 or, if appropriate, write $\alpha = \vec{0}$.

   b. Determine whether α is positive (+), negative (−), or zero (0).

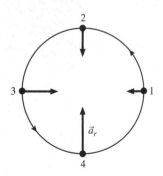
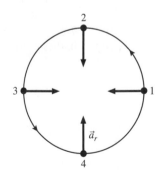
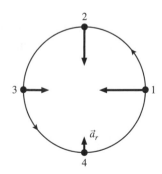

      $\alpha =$ _____ $\alpha =$ _____ $\alpha =$ _____

9 Impulse and Momentum

9.1 Momentum and Impulse

9.2 Solving Impulse and Momentum Problems

1. Rank in order, from largest to smallest, the momenta $(p_x)_1$ to $(p_x)_5$.

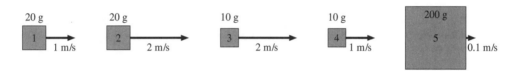

Order:

2. The position-versus-time graph is shown for a 500 g object. Draw the corresponding momentum-versus-time graph. Supply an appropriate scale on the vertical axis.

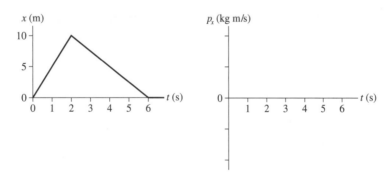

3. The momentum-versus-time graph is shown for a 500 g object. Draw the corresponding acceleration-versus-time graph. Supply an appropriate scale on the vertical axis.

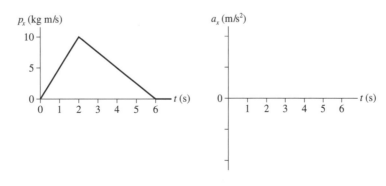

4. A 2 kg object is moving to the right with a speed of 1 m/s when it experiences an impulse due to the force shown in the graph. What is the object's speed and direction after the impulse?

a. F_x (N)

b. F_x (N)

5. A 2 kg object is moving to the left with a speed of 1 m/s when it experiences an impulse due to the force shown in the graph. What is the object's speed and direction after the impulse?

a. F_x (N)

b. F_x (N)

6. A 2 kg object has the velocity graph shown.

a. What is the object's initial momentum? _____

b. What is the object's final momentum? _____

c. What impulse does the object experience? _____

d. Draw the graph showing the force on the object.

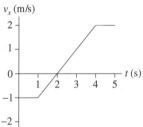

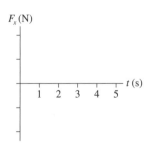

7. A carnival game requires you to knock over a wood post by throwing a ball at it. You're offered a very bouncy rubber ball and a very sticky clay ball of equal mass. Assume that you can throw them with equal speed and equal accuracy. You only get one throw.

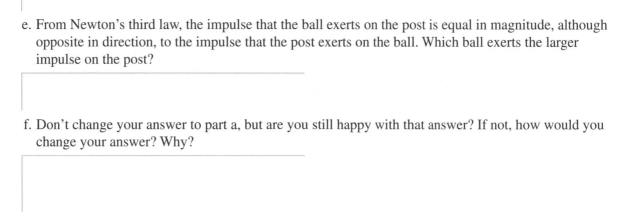

a. Which ball will you choose? Why?

b. Let's think about the situation more carefully. Both balls have the same initial momentum p_{ix} just before hitting the post. The clay ball sticks, the rubber ball bounces off with essentially no loss of speed. In terms of p_{ix}, what is the final momentum of each ball?

Clay ball: $p_{fx} = $ _____   Rubber ball $p_{fx} = $ _____

Hint: Momentum has a sign. Did you take the sign into account?

c. What is the *change* in the momentum of each ball?

Clay ball: $\Delta p_x = $ _____   Rubber ball $\Delta p_x = $ _____

d. Which ball experiences a larger impulse during the collision? Explain.

e. From Newton's third law, the impulse that the ball exerts on the post is equal in magnitude, although opposite in direction, to the impulse that the post exerts on the ball. Which ball exerts the larger impulse on the post?

f. Don't change your answer to part a, but are you still happy with that answer? If not, how would you change your answer? Why?

8. A falling rubber ball bounces on the floor.

a. Use the language of force, acceleration, and action/reaction to describe what happens.

b. Use the language of impulse and momentum to describe what happens.

9. A small, light ball S and a large, heavy ball L move toward each other, collide, and bounce apart.

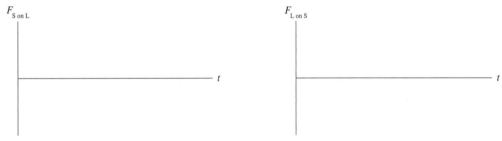

a. Compare the force that S exerts on L to the force that L exerts on S. That is, is $F_{S \, on \, L}$ larger, smaller, or equal to $F_{L \, on \, S}$? Explain.

b. Compare the time interval during which S experiences a force to the time interval during which L experiences a force. Are they equal, or is one longer than the other?

c. Sketch a graph showing a *plausible* $F_{S \, on \, L}$ as a function of time and another graph showing a plausible $F_{L \, on \, S}$ as a function of time. Be sure think about the *sign* of each force.

$F_{S \, on \, L}$

t

$F_{L \, on \, S}$

t

d. Compare the impulse delivered to S to the impulse delivered to L. Explain.

e. Compare the momentum change of S to the momentum change of L.

f. Compare the velocity change of S to the velocity change of L.

g. What is the change in the *sum* of the momenta of the two balls? Is it positive, negative, or zero?

© 2008 by Pearson Education, Inc., publishing as Pearson Addison-Wesley.

Exercises 10–12: Draw a momentum bar chart to show the momenta and impulse for the situation described.

10. A compressed spring shoots a ball to the right. The ball was initially at rest.

$$+ \quad p_{ix} \quad + \quad J_x \quad = \quad p_{fx}$$

$$0$$

$$-$$

11. A rubber ball is tossed straight up and bounces off the ceiling. Consider only the collision with the ceiling.

$$+ \quad p_{iy} \quad + \quad J_y \quad = \quad p_{fy}$$

$$0$$

$$-$$

12. A clay ball is tossed straight up and sticks to the ceiling. Consider only the collision with the ceiling.

$$+ \quad p_{iy} \quad + \quad J_y \quad = \quad p_{fy}$$

$$0$$

$$-$$

9.3 Conservation of Momentum

13. A golf club continues forward after hitting the golf ball. Is momentum conserved in the collision? Explain, making sure you are careful to identify the "system."

14. As you release a ball, it falls—gaining speed and momentum. Is momentum conserved?

    a. Answer this question from the perspective of choosing the ball alone as the system.

    b. Answer this question from the perspective of choosing ball + earth as the system.

15. Two particles collide, one of which was initially moving and the other initially at rest.

    a. Is it possible for *both* particles to be at rest after the collision? Give an example in which this happens, or explain why it can't happen.

    b. Is it possible for *one* particle to be at rest after the collision? Give an example in which this happens, or explain why it can't happen.

9.4 Inelastic Collisions

9.5 Explosions

Exercises 16–18: Prepare a pictorial representation for these problems, but *do not* solve them.
- Draw pictures of "before" and "after."
- Define symbols relevant to the problem.
- List known information, and identify the desired unknown.

16. A 50 kg archer, standing on frictionless ice, shoots a 100 g arrow at a speed of 100 m/s. What is the recoil speed of the archer?

17. The parking brake on a 2000 kg Cadillac has failed, and it is rolling slowly, at 1 mph, toward a group of small innocent children. As you see the situation, you realize there is just time for you to drive your 1000 kg Volkswagen head-on into the Cadillac and thus to save the children. With what speed should you impact the Cadillac to bring it to a halt?

18. Dan is gliding on his skateboard at 4 m/s. He suddenly jumps backward off the skateboard, kicking the skateboard forward at 8 m/s. How fast is Dan going as his feet hit the ground? Dan's mass is 50 kg and the skateboard's mass is 5 kg.

9.6 Momentum in Two Dimensions

19. An object initially at rest explodes into three fragments. The momentum vectors of two of the fragments are shown. Draw the momentum vector $\vec{p}_3$ of the third fragment.

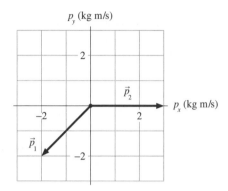

20. An object initially at rest explodes into three fragments. The momentum vectors of two of the fragments are shown. Draw the momentum vector $\vec{p}_3$ of the third fragment.

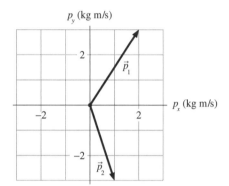

21. A 500 g ball traveling to the right at 8.0 m/s collides with and bounces off another ball. The figure shows the momentum vector $\vec{p}_1$ of the first ball after the collision. Draw the momentum vector $\vec{p}_2$ of the second ball.

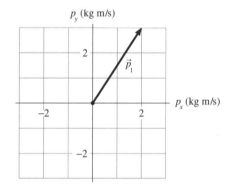

22. A 500 g ball traveling to the right at 4.0 m/s collides with and bounces off another ball. The figure shows the momentum vector $\vec{p}_1$ of the first ball after the collision. Draw the momentum vector $\vec{p}_2$ of the second ball.

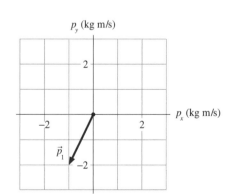

10 Energy

10.2 Kinetic Energy and Gravitational Potential Energy

10.3 A Closer Look at Gravitational Potential Energy

1. On the axes below, draw graphs of the kinetic energy of
   a. A 1000 kg car that uniformly accelerates from 0 to 20 m/s in 20 s.
   b. A 1000 kg car moving at 20 m/s that brakes to a halt with uniform deceleration in 20 s.
   c. A 1000 kg car that drives once around a 130-m-diameter circle at a speed of 20 m/s.

   Calculate K at several times, plot the points, and draw a smooth curve between them.

a.
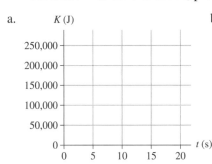
b.
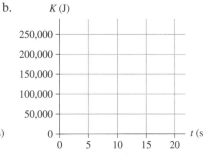
c.
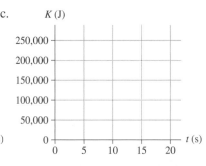

2. Below we see a 1 kg object that is initially 1 m above the ground and rises to a height of 2 m. Anjay, Brittany, and Carlos each measure its position, but each of them uses a different coordinate system. Fill in the table to show the initial and final gravitational potential energies and ΔU as measured by our three aspiring scientists.

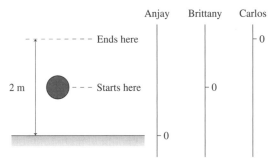

	U_i	U_f	ΔU
Anjay			
Brittany			
Carlos			

3. A roller coaster car rolls down a frictionless track, reaching speed v_f at the bottom.

   a. If you want the car to go twice as fast at the bottom, by what factor must you increase the height of the track?

   b. Does your answer to part a depend on whether the track is straight or not? Explain.

4. Below are shown three frictionless tracks. A ball is released from rest at the position shown on the left. To which point does the ball make it on the right before reversing direction and rolling back? Point B is the same height as the starting position.

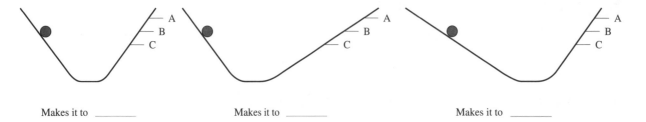

Makes it to _____   Makes it to _____   Makes it to _____

Exercises 5–7: Draw an energy bar chart to show the energy transformations for the situation described.

5. A car runs out of gas and coasts up a hill until finally stopping.

$$K_i \quad + \quad U_{gi} \quad = \quad K_f \quad + \quad U_{gf}$$

6. A pendulum is held out at 45° and released from rest. A short time later it swings through the lowest point on its arc.

$$K_i \quad + \quad U_{gi} \quad = \quad K_f \quad + \quad U_{gf}$$

7. A ball starts from rest on the top of one hill, rolls without friction through a valley, and just barely makes it to the top of an adjacent hill.

$$K_i \quad + \quad U_{gi} \quad = \quad K_f \quad + \quad U_{gf}$$

10.4 Restoring Forces and Hooke's Law

8. A spring is attached to the floor and pulled straight up by a string. The string's tension is measured. The graph shows the tension in the string as a function of the spring's length L.

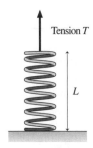 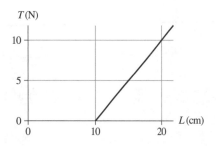

a. Does this spring obey Hooke's Law? Explain why or why not.

b. If it does, what is the spring constant?

9. Draw a figure analogous to Figure 10.16 in the textbook for a spring that is attached to a wall on the *right* end. Use the figure to show that F and Δs always have opposite signs.

10. A spring has an unstretched length of 10 cm. It exerts a restoring force F when stretched to a length of 11 cm.

   a. For what length of the spring is its restoring force $3F$?

   b. At what compressed length is the restoring force $2F$?

11. The left end of a spring is attached to a wall. When Bob pulls on the right end with a 200 N force, he stretches the spring by 20 cm. The same spring is then used for a tug-of-war between Bob and Carlos. Each pulls on his end of the spring with a 200 N force.

   a. How far does Bob's end of the spring move? Explain.

   b. How far does Carlos's end of the spring move? Explain.

10.5 Elastic Potential Energy

12. A heavy object is released from rest at position 1 above a spring. It falls and contacts the spring at position 2. The spring achieves maxiumum compression at position 3. Fill in the table below to indicate whether each of the quantities are +, –, or 0 during the intervals 1→2, 2→3, and 1→3.

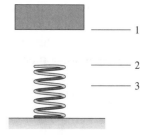

	1→2	2→3	1→3
ΔK			
ΔU_g			
ΔU_s			

13. Rank in order, from most to least, the amount of elastic potential energy $(U_\text{s})_1$ to $(U_\text{s})_4$ stored in each of these springs.

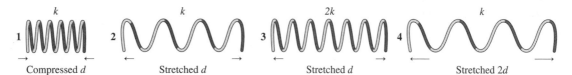

Order:

Explanation:

14. A spring gun shoots out a plastic ball at speed v_0. The spring is then compressed twice the distance it was on the first shot.

   a. By what factor is the spring's potential energy increased?

   b. By what factor is the ball's launch speed increased? Explain.

Exercises 15–16: Draw an energy bar chart to show the energy transformations for the situation described.

15. A bobsled sliding across frictionless, horizontal ice runs into a giant spring. A short time later the spring reaches its maximum compression.

$$K_i \quad + \quad U_{gi} \quad + \quad U_{si} \quad = \quad K_f \quad + \quad U_{gf} \quad + \quad U_{sf}$$

16. A brick is held above a spring that is standing on the ground. The brick is released from rest, and a short time later the spring reaches its maximum compression.

$$K_i \quad + \quad U_{gi} \quad + \quad U_{si} \quad = \quad K_f \quad + \quad U_{gf} \quad + \quad U_{sf}$$

10.6 Elastic Collisions

17. Ball 1 with an initial speed of 14 m/s has a perfectly elastic collision with ball 2 that is initially at rest. Afterward, the speed of ball 2 is 21 m/s.

   a. What will be the speed of ball 2 if the initial speed of ball 1 is doubled?

   b. What will be the speed of ball 2 if the mass of ball 1 is doubled?

10.7 Energy Diagrams

18. The figure shows a potential-energy curve. Suppose a particle with total energy E_1 is at position A and moving to the right.

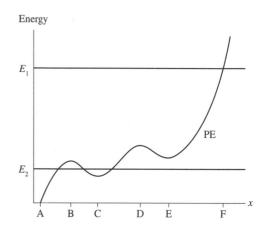

a. For each of the following regions of the x-axis, does the particle speed up, slow down, maintain a steady speed, or change direction?

A to B _____

B to C _____

C to D _____

D to E _____

E to F _____

b. Where is the particle's turning point? _____

c. For a particle that has total energy E_2, what are the possible motions and where do they occur along the x-axis?

d. What position or positions are points of stable equilibrium? For each, would a particle in equilibrium at that point have total energy $\leq E_2$, between E_2 and E_1, or $\geq E_1$?

e. What position or positions are points of unstable equilibrium? For each, would a particle in equilibrium at that point have total energy $\leq E_2$, between E_2 and E_1, or $\geq E_1$?

19. A particle with the potential energy shown in the graph is moving to the right at $x = 0$ m with total energy E.

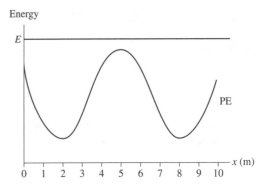

Energy

   a. At what value or values of x is the particle's speed a maximum?

   b. At what value or values of x is the particle's speed a minimum?

   c. At what value or values of x is the potential energy a maximum?

   d. Does this particle have a turning point in the range of x covered by the graph? If so, where?

20. Below are a set of axes on which you are going to draw a potential-energy curve. By doing experiments, you find the following information:
   • A particle with energy E_1 oscillates between positions D and E.
   • A particle with energy E_2 oscillates between positions C and F.
   • A particle with energy E_3 oscillates between positions B and G.
   • A particle with energy E_4 enters from the right, bounces at A, then never returns.

   Draw a potential-energy curve that is consistent with this information.

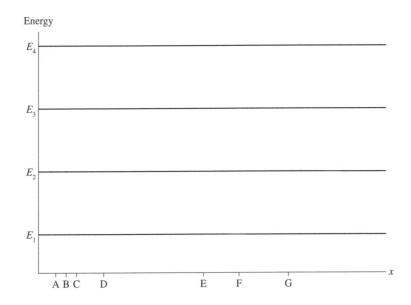

11 Work

11.2 Work and Kinetic Energy

11.3 Calculating and Using Work

1. For each pair of vectors, is the sign of $\vec{A} \cdot \vec{B}$ positive (+), negative (−), or zero (0)?

a.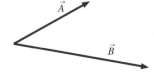

Sign = _____

b.

Sign = _____

c.

Sign = _____

d.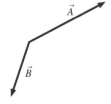

Sign = _____

e.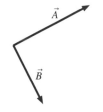

Sign = _____

f.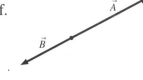

Sign = _____

2. Each of the diagrams below shows a vector $\vec{A}$. Draw and label a vector $\vec{B}$ that will cause $\vec{A} \cdot \vec{B}$ to have the sign indicated.

a.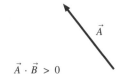

$\vec{A} \cdot \vec{B} > 0$

b.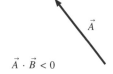

$\vec{A} \cdot \vec{B} < 0$

c.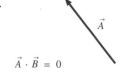

$\vec{A} \cdot \vec{B} = 0$

Exercises 3–10: For each situation:
- Draw a before-and-after pictorial diagram.
- Draw and label the displacement vector $\Delta\vec{r}$ on your diagram.
- Draw a free-body diagram showing *all* forces acting on the object.
- Make a table beside your diagrams showing the sign (+, –, or 0) of (i) the work done by each force seen in your free-body diagram, (ii) the net work W_{net}, and (iii) ΔK, the object's change in kinetic energy.

3. An elevator moves upward at constant speed.

4. A descending elevator brakes to a halt.

5. A box slides down a frictionless slope.

6. A box slides up a frictionless slope.

7. A ball is thrown straight up. Consider the ball from one microsecond after it leaves your hand until the highest point of its trajectory.

8. You toss a ball straight up. Consider the ball from the instant you begin moving your hand until you release the ball.

9. A car turns a corner at constant speed.

10. A flat block on a string swings once around a horizontal circle on a frictionless table. The block moves at steady speed.

11. A 0.2 kg plastic cart and a 20 kg lead cart both roll without friction on a horizontal surface. Equal forces are used to push both carts forward a distance of 1 m, starting from rest. After traveling 1 m, is the kinetic energy of the plastic cart greater than, less than, or equal to the kinetic energy of the lead cart? Explain.

12. Particle A has less mass than particle B. Both are pushed forward across a frictionless surface by equal forces for 1 s. Both start from rest.

   a. Compare the amount of work done on each particle. That is, is the work done on A greater than, less than, or equal to the work done on B? Explain.

   b. Compare the impulses delivered to particles A and B. Explain.

   c. Compare the final speeds of particles A and B. Explain.

11.4 The Work Done by a Variable Force

13. In Chapter 9, we found a graphical interpretation of Δp as the area under the F-versus-t graph from an initial time t_i to a final time t_f. Provide an analogous graphical interpretation of ΔK, the change in kinetic energy.

14. A particle moving along the x-axis experiences the forces shown below. How much work does each force do on the particle? What is each particle's change in kinetic energy?

a.

   $W =$ _____

   $\Delta K =$ _____

b.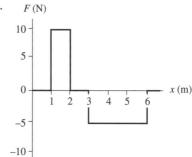

   $W =$ _____

   $\Delta K =$ _____

15. A 1 kg particle moving along the x-axis experiences the force shown in the graph. If the particle's speed is 2 m/s at $x = 0$ m, what is its speed when it gets to $x = 5$ m?

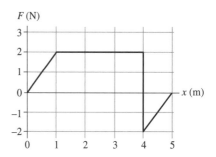

16. In Example 11.8 in the textbook, a compressed spring with a spring constant of 20 N/m expands from $x_0 = -20$ cm $= -0.20$ m to its equilibrium position at $x_1 = 0$ m.

    a. Graph the spring force F_{sp} from $x_1 = -0.20$ m to $x_2 = 0$ m.

    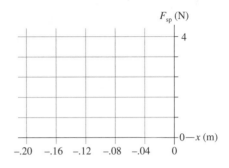

    b. Suppose the surface had been frictionless. Use your graph to determine ΔK, the change in a cube's kinetic energy when launched by a spring that has been compressed by 20 cm.

    c. Use your result from part b to find the launch speed of a 100 g cube in the absence of friction. Compare your answer to the value found in the Example 11.8. Why are they different?

11.5 Force, Work, and Potential Energy

11.6 Finding Force from Potential Energy

17. A particle moves in a vertical plane along a *closed* path, starting at A and eventually returning to its starting point. How much work is done on the particle by gravity? Explain.

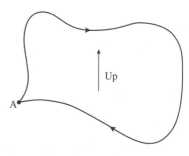

18. a. If the force on a particle at some point in space is zero, must its potential energy also be zero at that point? Explain.

    b. If the potential energy of a particle at some point in space is zero, must the force on it also be zero at that point? Explain.

19. The graph shows the potential-energy curve of a particle moving along the *x*-axis under the influence of a conservative force.

    a. In which intervals of *x* is the force on the particle to the right?

    b. In which intervals of *x* is the force on the particle to the left?

    c. At what value or values of *x* is the magnitude of the force a maximum?

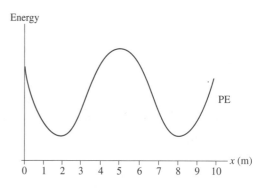

d. What value or values of x are positions of stable equilibrium?

e. What value or values of x are positions of unstable equilibrium?

f. If the particle is released from rest at $x = 0$ m, will it reach $x = 10$ m? Explain.

11.7 Thermal Energy

20. A ball of clay traveling at 10 m/s slams into a wall and sticks. What happened to the kinetic energy the clay had just before impact?

21. What energy transformations occur as a box slides up a gentle but slightly rough incline until stopping at the top?

11.8 Conservation of Energy

22. Give a *specific* example of a situation in which:

a. $W_{ext} \rightarrow K$ with $\Delta U = 0$ and $\Delta E_{th} = 0$.

b. $W_{ext} \rightarrow U$ with $\Delta K = 0$ and $\Delta E_{th} = 0$.

c. $K \rightarrow U$ with $W_{ext} = 0$ and $\Delta E_{th} = 0$.

d. $W_{ext} \rightarrow E_{th}$ with $\Delta K = 0$ and $\Delta U = 0$.

e. $U \rightarrow E_{th}$ with $\Delta K = 0$ and $W_{ext} = 0$.

23. A system loses 1000 J of potential energy. In the process, it does 500 J of work on the environment and the thermal energy increases by 250 J. Show this process on an energy bar chart.

$$K_i \quad + \quad U_i \quad + \quad W_{ext} \quad = \quad K_f \quad + \quad U_f \quad + \quad \Delta E_{th}$$

24. A system gains 1000 J of kinetic energy while losing 500 J of potential energy. The thermal energy increases by 250 J. Show this process on an energy bar chart.

$$K_i \quad + \quad U_i \quad + \quad W_{ext} \quad = \quad K_f \quad + \quad U_f \quad + \quad \Delta E_{th}$$

25. A box is sitting at the top of a ramp. An external force pushes the box down the ramp, causing it to slowly accelerate. Show this process on an energy bar chart.

11.9 Power

26. a. If you push an object 10 m with a 10 N force in the direction of motion, how much work do you do on it?

   b. How much power must you provide to push the object in 1 s? In 10 s? In 0.1 s?

12 Rotation of a Rigid Body

12.1 Rotational Motion

1. The following figures show a wheel rolling on a ramp. Determine the signs (+ or −) of the wheel's angular velocity and angular acceleration.

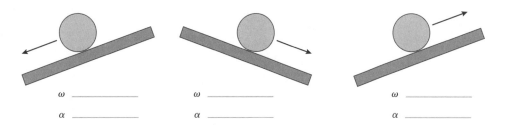

ω _____   ω _____   ω _____

α _____   α _____   α _____

2. A ball is rolling back and forth inside a bowl. The figure shows the ball at extreme left edge of the ball's motion as it changes direction.

   a. At this point, is ω positive, negative, or zero? _____
   b. At this point, is α positive, negative, or zero? _____

3. A wheel rolls to the left along a horizontal surface, up a ramp, then continues along the upper horizontal surface. Draw graphs for the wheel's angular velocity ω and angular acceleration α as a function of time.

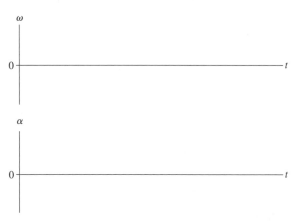

4. A wheel rolls to the right along the surface shown. Draw graphs for the wheel's angular velocity ω and angular acceleration α until the wheel reaches its highest point on the right side.

12.2 Rotation about the Center of Mass

5. Is the center of mass of this dumbbell at point 1, 2, or 3? Explain.

6. Mark the center of mass of this object with an ×.

12.3 Rotational Energy

12.4 Calculating the Moment of Inertia

7. The figure shows four equal-mass bars rotating about their center. Rank in order, from largest to smallest, their rotational kinetic energies K_1 to K_4.

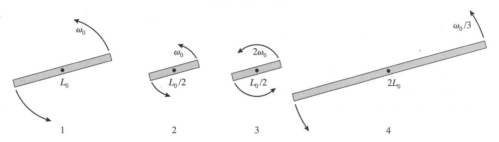

Order:

Explanation:

8. Two rotating solid spheres have the same angular velocity and the same mass. Sphere B has twice the rotational kinetic energy of sphere A.
   a. What is the ratio R_B/R_A of their radii?

   b. Would your answer differ if one sphere were solid and the other an equal-mass spherical shell? Explain.

9. Which has more kinetic energy: a particle of mass M rotating with angular velocity ω in a circle of radius R, or a sphere of mass M and radius R spinning at angular velocity ω? Explain.

10. The moment of inertia of a uniform rod about an axis through its center is $\frac{1}{12}ML^2$. The moment of inertia about an axis at one end is $\frac{1}{3}ML^2$. Explain *why* the moment of inertia is larger about the end than about the center.

11. You have two steel spheres. Sphere 2 has three times the radius of sphere 1. By what *factor* does the moment of inertia I_2 of sphere 2 exceed the moment of inertia I_1 of sphere 1?

12. The professor hands you two spheres. They have the same mass, the same radius, and the same exterior surface. The professor claims that one is a solid sphere and that the other is hollow. Can you determine which is which without cutting them open? If so, how? If not, why not?

13. Rank in order, from largest to smallest, the moments of inertia I_1, I_2, and I_3 about the midpoint of the rod.

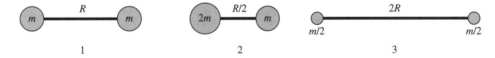

Order:

Explanation:

© 2008 by Pearson Education, Inc., publishing as Pearson Addison-Wesley.

12.5 Torque

14. Five forces are applied to a door. For each, determine if the torque about the hinge is positive (+), negative (−), or zero (0).

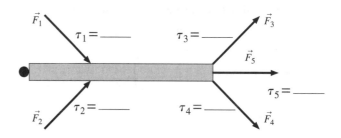

$\tau_1 =$ _____ $\tau_3 =$ _____

$\tau_2 =$ _____ $\tau_4 =$ _____

$\tau_5 =$ _____

15. Six forces, each of magnitude either F or $2F$, are applied to a door. Rank in order, from largest to smallest, the six torques τ_1 to τ_6 about the hinge.

Order:

Explanation:

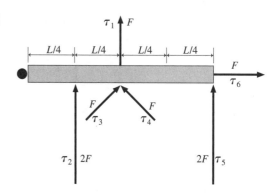

16. A bicycle is at rest on a smooth surface. A force is applied to the bottom pedal as shown. Does the bicycle roll forward (to the right), backward (to the left), or not at all? Explain.

17. Four forces are applied to a rod that can pivot on an axle. For each force,

    a. Use a **black** pen or pencil to draw the line of action.

    b. Use a **red** pen or pencil to draw and label the moment arm, or state that $d = 0$.

    c. Determine if the torque about the axle is positive (+), negative (−), or zero (0). Write your answer in the blank.

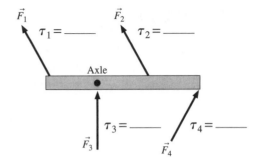

18. a. Draw a force vector at A whose torque about the axle is negative.

    b. Draw a force vector at B whose torque about the axle is zero.

    c. Draw a force vector at C whose torque about the axle is positive.

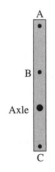

19. a. Draw a second force $\vec{F}_2$ that forms a couple with $\vec{F}_1$.

    b. Draw and label the distance l between their lines of action.

    c. Is the torque positive, negative, or zero? Explain.

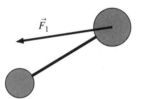

20. The dumbbells below are all the same size, and the forces all have the same magnitude. Rank in order, from largest to smallest, the torques τ_1, τ_2, and τ_3.

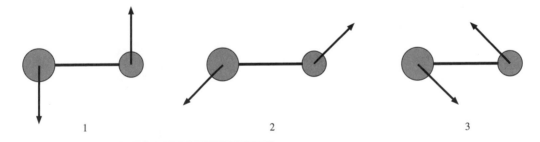

Order:

Explanation:

12.6 Rotational Dynamics

21. A student gives a quick push to a ball at the end of a massless, rigid rod, causing the ball to rotate clockwise in a *horizontal* circle. The rod's pivot is frictionless.

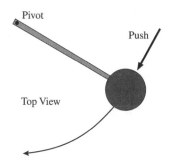

a. As the student is pushing, is the torque about the pivot positive, negative, or zero?

b. After the push has ended, does the ball's angular velocity

    i.   Steadily increase?

    ii.   Increase for awhile, then hold steady?

    iii. Hold steady?

    iv. Decrease for awhile, then hold steady?

    v.   Steadily decrease?

    Explain the reason for your choice.

c. Right after the push has ended, is the torque positive, negative, or zero? _____

22. a. Rank in order, from largest to smallest, the torques τ_1 to τ_4.

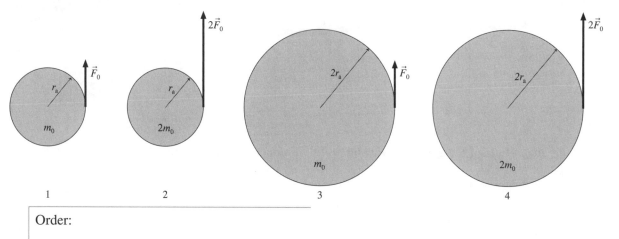

Order:

Explanation:

b. Rank in order, from largest to smallest, the angular accelerations α_1 to α_4.

23. The top graph shows the torque on a rotating wheel as a function of time. The wheel's moment of inertia is 10 kg m^2. Draw graphs of α-versus-t and ω-versus-t, assuming $\omega_0 = 0$. Provide units and appropriate scales on the vertical axes.

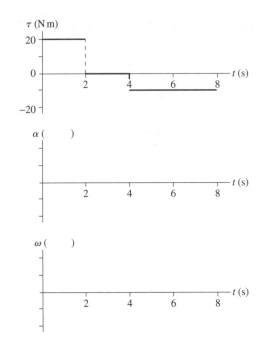

24. The wheel turns on a frictionless axle. A string wrapped around the smaller diameter shaft is tied to a block. The block is released at $t = 0$ s and hits the ground at $t = t_1$.

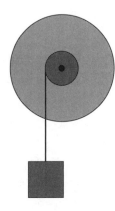

   a. Draw a graph of ω-versus-t for the wheel, starting at $t = 0$ s and continuing to some time $t > t_1$.

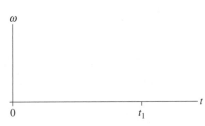

   b. Is the magnitude of the block's downward acceleration greater than g, less than g, or equal to g? Explain.

12.7 Rotation about a Fixed Axis

25. A square plate can rotate about an axle through its center. Four forces of equal magnitude are applied to different points on the plate. The forces turn as the plate rotates, maintaining the same orientation with respect to the plate. Rank in order, from largest to smallest, the angular accelerations α_1 to α_4.

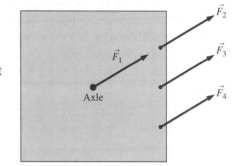
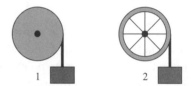

Order:

Explanation:

26. A solid cylinder and a cylindrical shell have the same mass, same radius, and turn on frictionless, horizontal axles. (The cylindrical shell has light-weight spokes connecting the shell to the axle.) A rope is wrapped around each cylinder and tied to a block. The blocks have the same mass and are held the same height above the ground. Both blocks are released simultaneously. The ropes do not slip.

    Which block hits the ground first? Or is it a tie? Explain.

12.8 Static Equilibrium

27. A uniform rod pivots about a frictionless, horizontal axle through its center. It is placed on a stand, held motionless in the position shown, then gently released. On the right side of the figure, draw the final, equilibrium position of the rod. Explain your reasoning.

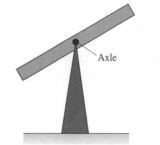

28. The dumbbell has masses m and $2m$. Force $\vec{F}_1$ acts on mass m in the direction shown. Is there a force $\vec{F}_2$ that can act on mass $2m$ such that the dumbbell moves with pure translational motion, without any rotation? If so, draw $\vec{F}_2$, making sure that its length shows the magnitude of $\vec{F}_2$ relative to $\vec{F}_1$. If not, explain why not.

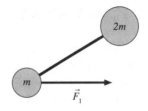

29. Forces $\vec{F}_1$ and $\vec{F}_2$ have the same magnitude and are applied to the corners of a square plate. Is there a *single* force $\vec{F}_3$ that, if applied to the appropriate point on the plate, will cause the plate to be in total equilibrium? If so, draw it, making sure it has the right position, orientation, and length. If not, explain why not.

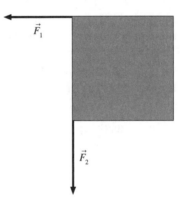

12.9 Rolling Motion

30. A wheel is rolling along a horizontal surface with the center-of-mass velocity shown. Draw the velocity vector $\vec{v}$ at points 1 to 4 on the rim of the wheel.

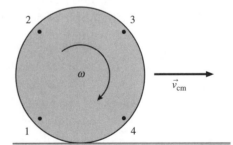

12.10 The Vector Description of Rotational Motion

12.11 Angular Momentum of a Rigid Body

31. For each vector pair $\vec{A}$ and $\vec{B}$ shown below, determine if $\vec{A} \times \vec{B}$ points into the page, out of the page, or is zero.

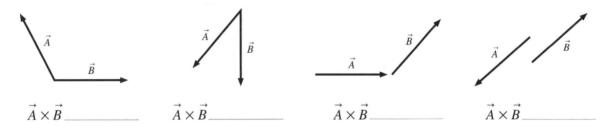

$\vec{A} \times \vec{B}$ _____ $\vec{A} \times \vec{B}$ _____ $\vec{A} \times \vec{B}$ _____ $\vec{A} \times \vec{B}$ _____

32. Each figure below shows $\vec{A}$ and $\vec{A} \times \vec{B}$. Determine if $\vec{B}$ is in the plane of the page or perpendicular to the page. If $\vec{B}$ is in the plane of the page, draw it. If $\vec{B}$ is perpendicular to the page, state whether $\vec{B}$ points into the page or out of the page.

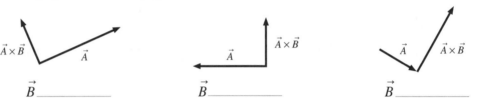

$\vec{B}$ _____ $\vec{B}$ _____ $\vec{B}$ _____

33. Draw the angular velocity vector on each of the rotating wheels.

a.   b.   c.

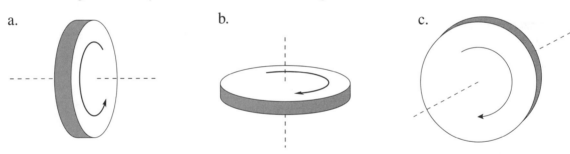

34. The figures below show a force acting on a particle. For each, draw the torque vector for the torque about the origin.
    • Place the tail of the torque vector at the origin.
    • Draw the vector large and straight (use a ruler!) so that its direction is clear. Use dotted lines from the tip of the vector to the axes to show the plane in which the vector lies.

a.   b.   c.

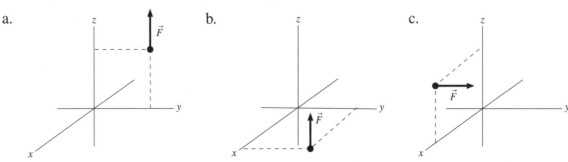

35. The figures below show a particle with velocity $\vec{v}$. For each, draw the angular momentum vector $\vec{L}$ for the angular momentum relative to the origin. Place the tail of the angular momentum vector at the origin.

a.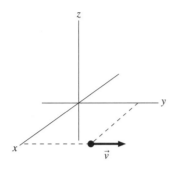

b.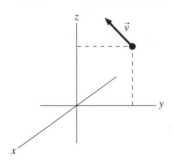

c.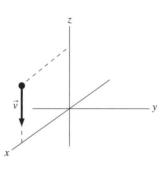

36. Rank in order, from largest to smallest, the angular momenta L_1 to L_4.

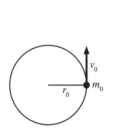

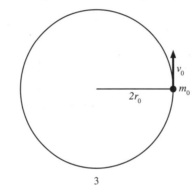

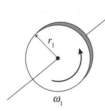

1   2   3   4

Order:

Explanation:

37. Disks 1 and 2 have equal mass. Is the angular momentum of disk 2 larger than, smaller than, or equal to the angular momentum of disk 1? Explain.

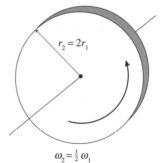

13 Newton's Theory of Gravity

13.1 A Little History

13.2 Isaac Newton

13.3 Newton's Law of Gravity

1. Is the earth's gravitational force on the moon larger than, smaller than, or equal to the moon's gravitational force on the earth? Explain.

2. Star A is twice as massive as star B. They attract each other.

    a. Draw gravitational force vectors on both stars. The length of each vector should be proportional to the size of the force.

     $m_A = 2m_B$ m_B

    b. Is the acceleration of star A larger than, smaller than, or equal to the acceleration of star B? Explain.

3. The gravitational force of a star on orbiting planet 1 is F_1. Planet 2, which is twice as massive as planet 1 and orbits at half the distance from the star, experiences gravitational force F_2. What is the ratio F_2/F_1?

4. Comets orbit the sun in highly elliptical orbits. A new comet is sighted at time t_1.

   a. Later, at time t_2, the comet's acceleration a_2 is twice as large as the acceleration a_1 it had at t_1. What is the ratio r_2/r_1 of the comet's distance from the sun at t_2 to its distance at t_1?

   Comet's orbit ⟶ Sun

   b. Still later, at time t_3, the comet has rounded the sun and is headed back out to the farthest reaches of the solar system. The size of the force F_3 on the comet at t_3 is the same as the size of force F_2 at t_2, but the comet's distance from the sun r_3 is only 90% of distance r_2. Astronomers recognize that the comet has lost mass. Part of it was "boiled away" by the heat of the sun during the time of closest approach, thus forming the comet's tail. What percent of its initial mass did the comet lose?

13.4 Little *g* and Big *G*

5. How far away from the earth does an orbiting spacecraft have to be in order for the astronauts inside to be weightless?

6. The free-fall acceleration at the surface of planet 1 is 20 m/s^2. The radius and the mass of planet 2 are half those of planet 1. What is *g* on planet 2?

13.5 Gravitational Potential Energy

7. Explain *why* the gravitational potential energy of two masses is negative. Note that saying "because that's what the formula gives" is *not* an explanation. An explanation makes use of the basic ideas of force and potential energy.

13.6 Satellite Orbits and Energies

8. Planet X orbits the star Alpha with a "year" that is 200 earth days long. Planet Y circles Alpha at nine times the distance of planet X. How long is a year on planet Y?

9. The mass of Jupiter is $M_{\text{Jupiter}} = 300M_{\text{earth}}$. Jupiter orbits around the sun with $T_{\text{Jupiter}} = 11.9$ years in an orbit with $r_{\text{Jupiter}} = 5.2r_{\text{earth}}$. Suppose the earth could be moved to the distance of Jupiter and placed in a circular orbit around the sun. The new period of the earth's orbit would be

a. 1 year.
b. 11.9 years.
c. Between 1 year and 11.9 years.
d. More than 11.9 years.
e. It could be anything, depending on the speed the earth is given.
f. It is impossible for a planet of earth's mass to orbit at the distance of Jupiter.

Circle the letter of the true statement. Then explain your choice.

10. Satellite A orbits a planet with a speed of 10,000 m/s. Satellite B is twice as massive as satellite A and orbits at twice the distance from the center of the planet. What is the speed of satellite B?

11. a. A crew of a spacecraft in a clockwise circular orbit around the moon wants to change to a new orbit that will take them down to the surface. In which direction should they fire the rocket engine? On the figure, show the exhaust gases coming out of the spacecraft.

    b. On the figure, show the spacecraft's orbit after firing its rocket engine.

    c. The moon has no atmosphere, so the spacecraft will continue unimpeded along its new orbit until either firing its rocket again or (ouch!) intersecting the surface. As it descends, does its speed increase, decrease, or stay the same? Explain.

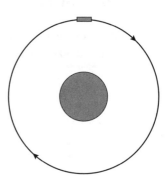

14 Oscillations

14.1 Simple Harmonic Motion

1. Give three examples of *oscillatory* motion. (Note that circular motion is not the same as oscillatory motion.)

2. On the axes below, sketch three cycles of the displacement-versus-time graph for:

   a. A particle undergoing symmetric periodic motion that is *not* SHM.

   b. A particle undergoing asymmetric periodic motion.

   c. A particle undergoing simple harmonic motion.

3. Consider the particle whose motion is represented by the *x*-versus-*t* graph below.

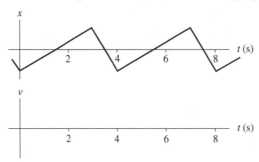

a. Is this periodic motion? _____ b. Is this motion SHM? _____

c. What is the period? _____ d. What is the frequency? _____

e. You learned in Chapter 2 to relate velocity graphs to position graphs. Use that knowledge to draw the particle's velocity-versus-time graph on the axes provided.

4. Shown below is the velocity-versus-time graph of a particle.

a. What is the period of the motion? _____

b. Draw the particle's position-versus-time graph, starting from *x* = 0 at *t* = 0 s.

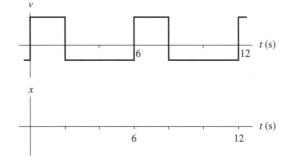

5. The figure shows the position-versus-time graph of a particle in SHM.

a. At what times is the particle moving to the right at maximum speed?

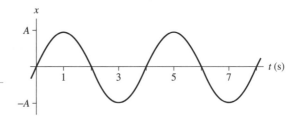

b. At what times is the particle moving to the left at maximum speed?

c. At what times is the particle instantaneously at rest?

14.2 Simple Harmonic Motion and Circular Motion

6. A particle goes around a circle 5 times at constant speed, taking a total of 2.5 seconds.

   a. Through what angle *in degrees* has the particle moved? _____

   b. Through what angle *in radians* has the particle moved? _____

   c. What is the particle's frequency f? _____

   d. Use your answer to part b to determine the particle's angular frequency ω.

   e. Does ω (in rad/s) $= 2\pi f$ (in Hz)? _____

7. A particle moves counterclockwise around a circle at constant speed. For each of the phase constants given below:
   - Show with a dot *on the circle* the particle's starting position.
   - Sketch two cycles of the particle's *x*-versus-*t* graph.

a.

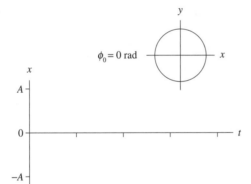

b.

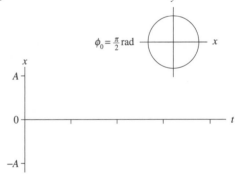

c.

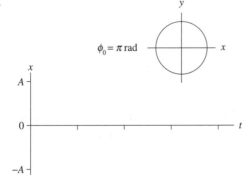

d.

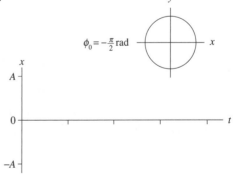

8. a. On the top set of axes below, sketch two cycles of the *x*-versus-*t* graphs for a particle in simple harmonic motion with phase constants i) $\phi_0 = \pi/2$ rad and ii) $\phi_0 = -\pi/2$ rad.

   b. Use the bottom set of axes to sketch velocity-versus-time graphs for the particles. Make sure each velocity graph aligns vertically with the correct points on the *x*-versus-*t* graph.

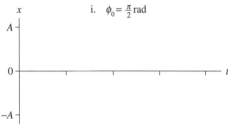
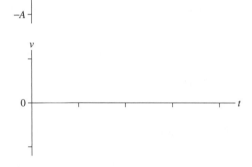

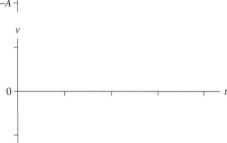

9. The graph below represents a particle in simple harmonic motion.

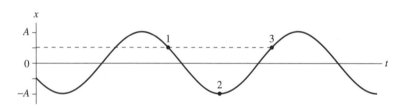
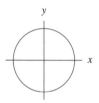

   a. What is the phase constant ϕ_0? Explain how you determined it.

   b. What is the phase of the particle at each of the three numbered points on the graph?

   Phase at 1: _____   Phase at 2: _____   Phase at 3: _____

   c. Place dots on the circle above to show the position of a circular-motion particle at the times corresponding to points 1, 2, and 3. Label each dot with the appropriate number.

14.3 Energy in Simple Harmonic Motion

10. The figure shows the potential-energy diagram of a particle oscillating on a spring.

    a. What is the spring's equilibrium length?

    b. The particle's turning points are at 14 cm and 26 cm. Draw the total energy line and label it TE.

    c. What is the particle's maximum kinetic energy?

    d. Draw a graph of the particle's kinetic energy as a function of position.

    e. What will be the turning points if the particle's total energy is doubled?

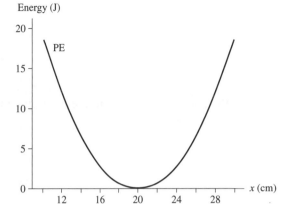

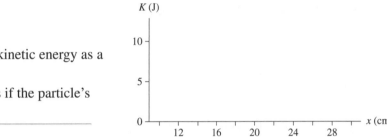

11. A block oscillating on a spring has an amplitude of 20 cm. What will be the block's amplitude if its total energy is tripled? Explain.

12. A block oscillating on a spring has a maximum speed of 20 cm/s. What will be the block's maximum speed if its total energy is tripled? Explain.

13. The figure shows the potential energy diagram of a particle.

    a. Is the particle's motion periodic? How can you tell?

    b. Is the particle's motion simple harmonic motion? How can you tell?

    c. What is the amplitude of the motion?

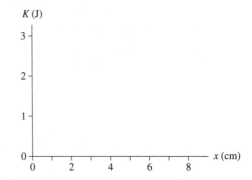

    d. Draw a graph of the particle's kinetic energy as a function of position.

14. Equation 14.25 in the textbook states that $\frac{1}{2}kA^2 = \frac{1}{2}mv_{max}^2$. What does this mean? Write a couple of sentences explaining how to interpret this equation.

14.4 The Dynamics of Simple Harmonic Motion

14.5 Vertical Oscillations

15. A block oscillating on a spring has period $T = 4$ s.

   a. What is the period if the block's mass is halved? Explain.
      Note: You do not know values for either m or k. Do *not* assume any particular values for them. The required analysis involves thinking about ratios.

   b. What is the period if the value of the spring constant is quadrupled?

   c. What is the period if the oscillation amplitude is doubled while m and k are unchanged?

16. For graphs a and b, determine:
   • The angular frequency ω.
   • The oscillation amplitude A.
   • The phase constant ϕ_0.

   Note: Graphs a and b are independent. Graph b is *not* the velocity graph of a.

a.

$\omega =$ _____

$A =$ _____

$\phi_0 =$ _____

b.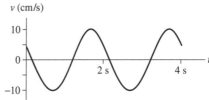

$\omega =$ _____

$A =$ _____

$\phi_0 =$ _____

17. The graph on the right is the position-versus-time graph for a simple harmonic oscillator.

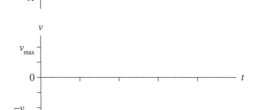

   a. Draw the v-versus-t and a-versus-t graphs.

   b. When x is greater than zero, is a ever greater than zero? If so, at which points in the cycle?

   c. When x is less than zero, is a ever less than zero? If so, at which points in the cycle?

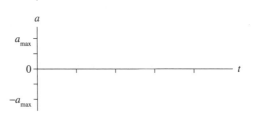

   d. Can you make a general conclusion about the relationship between the sign of x and the sign of a?

   e. When x is greater than zero, is v ever greater than zero? If so, how is the oscillator moving at those times?

18. For the oscillation shown on the left below:

   a. What is the phase constant ϕ_0? _____

   b. Draw the corresponding v-versus-t graph on the axes below the x-versus-t graph.

   c. On the axes on the right, sketch two cycles of the x-versus-t and the v-versus-t graphs if the value of ϕ_0 found in part a is replaced by its negative, $-\phi_0$.

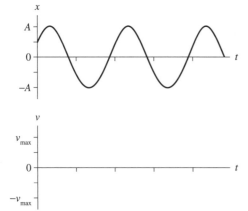
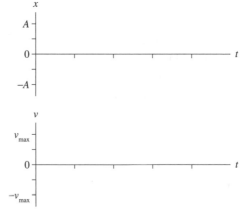

   d. Describe *physically* what is the same and what is different about the initial conditions for two oscillators having "equal but opposite" phase constants ϕ_0 and $-\phi_0$.

19. The top graph shows the position versus time for a mass oscillating on a spring. On the axes below, sketch the position-versus-time graph for this block for the following situations:

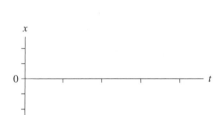

Note: The changes described in each part refer back to the original oscillation, not to the oscillation of the previous part of the question. Assume that all other parameters remain constant. Use the same horizontal and vertical scales as the original oscillation graph.

a. The amplitude and the frequency are doubled.

b. The amplitude is halved and the mass is quadrupled.

c. The phase constant is increased by $\pi/2$ rad.

d. The maximum speed is doubled while the amplitude remains constant.

14.6 The Pendulum

20. A pendulum on planet X, where the value of g is unknown, oscillates with a period of 2 seconds. What is the period of this pendulum if:

    a. Its mass is tripled?
       Note: You do not know the values of m, L, or g, so do not assume any specific values.

    b. Its length is tripled?

    c. Its oscillation amplitude is tripled?

21. The graph shows the displacement s versus time for an oscillating pendulum.

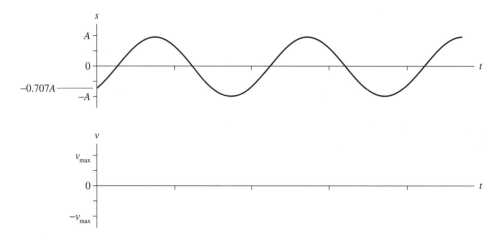

    a. Draw the pendulum's velocity-versus-time graph.
    b. What is the value of the phase constant ϕ_0?

    c. In the space at the right, draw a *picture* of the
       pendulum that shows (and labels!)
       • The extremes of its motion.
       • Its position at $t = 0$ s.
       • Its direction of motion (using an arrow) at $t = 0$ s.

14.7 Damped Oscillations

22. If the damping constant b of an oscillator is increased,

    a. Is the medium more resistive or less resistive? _____

    b. Do the oscillations damp out more quickly or less quickly? _____

    c. Is the time constant τ increased or decreased? _____

23. A block on a spring oscillates horizontally on a table with friction. Draw and label force vectors on the block to show all *horizontal* forces on the block.

    a. The mass is to the right of the equilibrium point and moving away from it.

    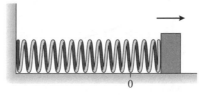

    b . The mass is to the right of the equilibrium point and approaching it.

    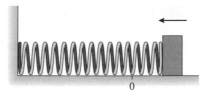

24. A mass oscillating on a spring has a frequency of 0.5 Hz and a damping time constant $\tau = 5$ s. Use the grid below to draw a reasonably accurate position-versus-time graph lasting 40 s.

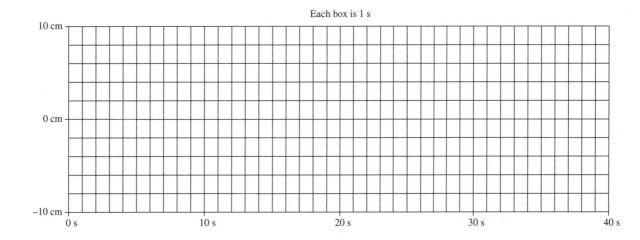

25. The figure below shows the envelope of the oscillations of a damped oscillator. On the same axes, draw the envelope of oscillations if

   a. The time constant is doubled.

   b. The time constant is halved.

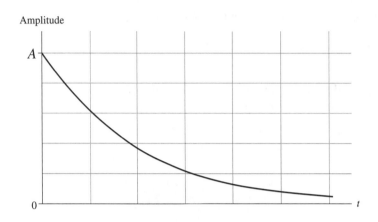

14.8 Driven Oscillations and Resonance

26. What is the difference between the driving frequency and the natural frequency of an oscillator?

27. A car drives along a bumpy road on which the bumps are equally spaced. At a speed of 20 mph, the frequency of hitting bumps is equal to the natural frequency of the car bouncing on its springs.

   a. Draw a graph of the car's vertical bouncing amplitude as a function of its speed if the car has new shock absorbers (large damping coefficient).

   b. Draw a graph of the car's vertical bouncing amplitude as a function of its speed if the car has worn out shock absorbers (small damping coefficient).

   Draw both graphs on the same axes, and label them as to which is which.

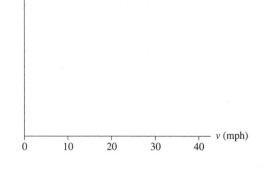

15 Fluids and Elasticity

15.1 Fluids

1. An object has density ρ.

   a. Suppose each of the object's three dimensions is increased by a factor of 2 without changing the material of which the object is made. Will the density change? If so, by what factor? Explain.

   b. Suppose each of the object's three dimensions is increased by a factor of 2 without changing the object's mass. Will the density change? If so, by what factor? Explain.

2. Air enclosed in a cylinder has density $\rho = 1.4$ kg/m^3.

   a. What will be the density of the air if the length of the cylinder is doubled while the radius is unchanged?

   b. What will be the density of the air if the radius of the cylinder is halved while the length is unchanged?

3. Air enclosed in a sphere has density $\rho = 1.4$ kg/m^3. What will the density be if the radius of the sphere is halved?

© 2008 by Pearson Education, Inc., publishing as Pearson Addison-Wesley.

15.2 Pressure

15.3 Measuring and Using Pressure

4. When you stand on a bathroom scale, it reads 700 N. Suppose a giant vacuum cleaner sucks half the air out of the room, reducing the pressure to 0.5 atm. Would the scale reading increase, decrease, or stay the same? Explain.

5. Rank in order, from largest to smallest, the pressures at A, B, and C.

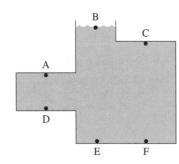

Order:

Explanation:

6. Refer to the figure in Exercise 5. Rank in order, from largest to smallest, the pressures at D, E, and F.

Order:

Explanation:

7. The gauge pressure at the bottom of a cylinder of liquid is $p_g = 0.4$ atm. The liquid is poured into another cylinder with twice the radius of the first cylinder. What is the gauge pressure at the bottom of the second cylinder?

8. Cylinders A and B contain liquids. The pressure p_A at the bottom of A is higher than the pressure p_B at the bottom of B. Is the ratio p_A/p_B of the absolute pressures larger than, smaller than, or equal to the ratio of the gauge pressures? Explain.

9. A and B are rectangular tanks full of water. They have equal depths, equal thicknesses (the dimension into the page), but different widths.

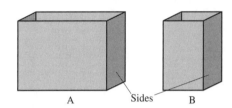

A   Sides   B

   a. Compare the forces the water exerts on the bottoms of the tanks. Is F_A larger than, smaller than, or equal to F_B? Explain.

   b. Compare the forces the water exerts on the sides of the tanks. Is F_A larger than, smaller than, or equal to F_B? Explain.

10. Water expands when heated. Suppose a beaker of water is heated from 10°C to 90°C. Does the pressure at the bottom of the beaker increase, decrease, or stay the same? Explain.

11. Is p_A larger than, smaller than, or equal to p_B? Explain.

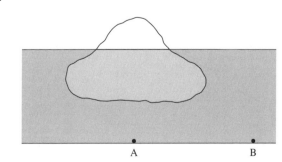

A   B

12. The container shown holds a mixture of oil and water. To begin, the container is shaken vigorously to mix the oil into the water by breaking it into very tiny droplets. This is what happens when you shake a jar of salad dressing. Eventually, the oil separates and rises to the top. Oil and water are *immiscible*, meaning that the total volume is the same whether they are mixed or separated. The pressure at the bottom of the container after the oil has separated is *not* the same as the initial pressure when the oil and water are mixed, although it may take some careful thought to understand why.

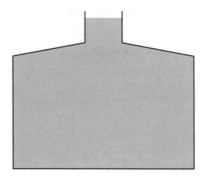

Is the final pressure at the bottom higher or lower than the initial pressure? Explain.

13. At sea level, the height of the mercury column in a sealed glass tube is 380 mm. What can you say about the contents of the space above the mercury? Be as specific as you can.

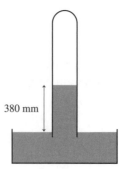

380 mm

15.4 Buoyancy

14. Rank in order, from largest to smallest, the densities of A, B, and C.

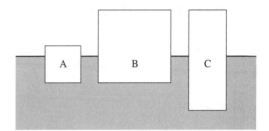

Order:

Explanation:

15. A, B, and C have the same volume. Rank in order, from largest to smallest, the sizes of the buoyant forces F_A, F_B, and F_C on A, B, and C.

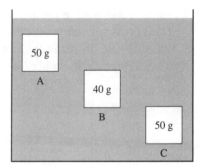

Order:

Explanation:

16. Refer to the figure of Exercise 15. Now A, B, and C have the same density. Rank in order, from largest to smallest, the sizes of the buoyant forces on A, B, and C.

Order:

Explanation:

17. Suppose you stand on a bathroom scale that is on the bottom of a swimming pool. The water comes up to your waist.
    Is the scale reading your weight? If not, does the scale read more than or less than your weight? Explain.

18. Ships A and B have the same height and the same mass. Their cross section profiles are shown in the figure. Does one ship ride higher in the water (more height above the water line) than the other? If so, which one? Explain.

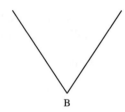

A   B

15.5 Fluid Dynamics

19. Gas flows through a pipe. You can't see into the pipe to know how the inner diameter changes. Rank in order, from largest to smallest, the gas speeds v_1 to v_3 at points 1, 2, and 3.

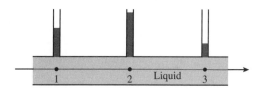

Order:

Explanation:

20. Liquid flows through a pipe. You can't see into the pipe to know how the inner diameter changes. Rank in order, from largest to smallest, the flow speeds v_1 to v_3 at points 1, 2, and 3.

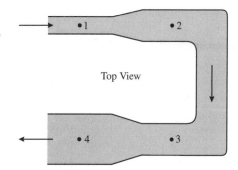

Order:

Explanation:

21. Liquid flows through this pipe. This is an overhead view.

a. Rank in order, from largest to smallest, the flow speeds v_1 to v_4 at points 1 to 4.

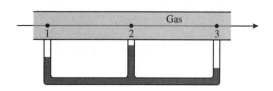

Order:

Explanation:

b. Rank in order, from largest to smallest, the pressures p_1 to p_4 at points 1 to 4.

Order:

Explanation:

22. Wind blows over a house. A window on the ground floor is open. Is there an air flow through the house? If so, does the air flow in the window and out the chimney, or in the chimney and out the window? Explain.

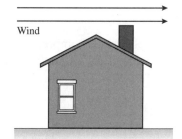

15.6 Elasticity

23. A force stretches a wire by 1 mm.

   a. A second wire of the same material has the same cross section and twice the length. How far will it be stretched by the same force? Explain.

   b. A third wire of the same material has the same length and twice the diameter as the first. How far will it be stretched by the same force? Explain.

24. A 2000 N force stretches a wire by 1 mm.

   a. A second wire of the same material is twice as long and has twice the diameter. How much force is needed to stretch it by 1 mm? Explain.

   b. A third wire is twice as long as the first and has the same diameter. How far is it stretched by a 4000 N force?

25. A wire is stretched right to the breaking point by a 5000 N force. A longer wire made of the same material has the same diameter. Is the force that will stretch it right to the breaking point larger than, smaller than, or equal to 5000 N? Explain.

26. Sphere A is compressed by 1% at an ocean depth of 4000 m. Sphere B is compressed by 1% at an ocean depth of 5000 m. Which has the larger bulk modulus? Explain.

DYNAMICS WORKSHEET Name _____ Problem _____

MODEL Make simplifying assumptions.

VISUALIZE

- Draw a picture. Show important points in the motion.
- Establish a coordinate system. Define symbols.
- List knowns. Identify what you're trying to find.

- Draw a motion diagram.
- Identify forces and interactions.
- Draw free-body diagrams.

Known

Find

SOLVE

Start with Newton's first or second law in component form, adding other information as needed to solve the problem.

ASSESS

Have you answered the question?

Do you have correct units, signs, and significant figures?

Is your answer reasonable?

DYNAMICS WORKSHEET Name _____ Problem _____

MODEL Make simplifying assumptions.

VISUALIZE

- Draw a picture. Show important points in the motion.
- Establish a coordinate system. Define symbols.
- List knowns. Identify what you're trying to find.

- Draw a motion diagram.
- Identify forces and interactions.
- Draw free-body diagrams.

Known

Find

SOLVE
Start with Newton's first or second law in component form, adding other information as needed to solve the problem.

ASSESS
Have you answered the question?
Do you have correct units, signs, and significant figures?
Is your answer reasonable?

DYNAMICS WORKSHEET   Name _____   Problem _____

MODEL   Make simplifying assumptions.

VISUALIZE

- Draw a picture. Show important points in the motion.
- Establish a coordinate system. Define symbols.
- List knowns. Identify what you're trying to find.

- Draw a motion diagram.
- Identify forces and interactions.
- Draw free-body diagrams.

Known

Find

SOLVE

Start with Newton's first or second law in component form, adding other information as needed to solve the problem.

ASSESS

Have you answered the question?

Do you have correct units, signs, and significant figures?

Is your answer reasonable?

DYNAMICS WORKSHEET   Name _____ Problem _____

MODEL Make simplifying assumptions.

VISUALIZE

- Draw a picture. Show important points in the motion.
- Establish a coordinate system. Define symbols.
- List knowns. Identify what you're trying to find.
- Draw a motion diagram.
- Identify forces and interactions.
- Draw free-body diagrams.

Known

Find

SOLVE
Start with Newton's first or second law in component form, adding other information as needed to solve the problem.

ASSESS
Have you answered the question?
Do you have correct units, signs, and significant figures?
Is your answer reasonable?

DYNAMICS WORKSHEET Name _____ Problem _____

VISUALIZE

- Draw a picture. Show important points in the motion.
- Establish a coordinate system. Define symbols.
- List knowns. Identify what you're trying to find.
- Draw a motion diagram.
- Identify forces and interactions.
- Draw free-body diagrams.

Known

Find

SOLVE

Start with Newton's first or second law in component form, adding other information as needed to solve the problem.

ASSESS

Have you answered the question?

Do you have correct units, signs, and significant figures?

Is your answer reasonable?

DYNAMICS WORKSHEET Name _____ Problem _____

MODEL Make simplifying assumptions.

VISUALIZE
- Draw a picture. Show important points in the motion.
- Establish a coordinate system. Define symbols.
- List knowns. Identify what you're trying to find.
- Draw a motion diagram.
- Identify forces and interactions.
- Draw free-body diagrams.

Known

Find

SOLVE
Start with Newton's first or second law in component form, adding other information as needed to solve the problem.

ASSESS
Have you answered the question?
Do you have correct units, signs, and significant figures?
Is your answer reasonable?

DYNAMICS WORKSHEET Name _____ Problem _____

VISUALIZE

- Draw a picture. Show important points in the motion.
- Establish a coordinate system. Define symbols.
- List knowns. Identify what you're trying to find.
- Draw a motion diagram.
- Identify forces and interactions.
- Draw free-body diagrams.

Known

Find

SOLVE

Start with Newton's first or second law in component form, adding other information as needed to solve the problem.

ASSESS

Have you answered the question?

Do you have correct units, signs, and significant figures?

Is your answer reasonable?

DYNAMICS WORKSHEET   Name _____   Problem _____

MODEL Make simplifying assumptions.

VISUALIZE

- Draw a picture. Show important points in the motion.
- Establish a coordinate system. Define symbols.
- List knowns. Identify what you're trying to find.

- Draw a motion diagram.
- Identify forces and interactions.
- Draw free-body diagrams.

Known

Find

SOLVE

Start with Newton's first or second law in component form, adding other information as needed to solve the problem.

ASSESS

Have you answered the question?

Do you have correct units, signs, and significant figures?

Is your answer reasonable?

DYNAMICS WORKSHEET Name _____ Problem _____

MODEL Make simplifying assumptions.

VISUALIZE

- Draw a picture. Show important points in the motion.
- Establish a coordinate system. Define symbols.
- List knowns. Identify what you're trying to find.

- Draw a motion diagram.
- Identify forces and interactions.
- Draw free-body diagrams.

Known

Find

SOLVE

Start with Newton's first or second law in component form, adding other information as needed to solve the problem.

ASSESS

Have you answered the question?
Do you have correct units, signs, and significant figures?
Is your answer reasonable?

DYNAMICS WORKSHEET Name _____ Problem _____

MODEL Make simplifying assumptions.

VISUALIZE

- Draw a picture. Show important points in the motion.
- Establish a coordinate system. Define symbols.
- List knowns. Identify what you're trying to find.
- Draw a motion diagram.
- Identify forces and interactions.
- Draw free-body diagrams.

Known

Find

SOLVE

Start with Newton's first or second law in component form, adding other information as needed to solve the problem.

ASSESS

Have you answered the question?

Do you have correct units, signs, and significant figures?

Is your answer reasonable?

DYNAMICS WORKSHEET   Name _____   Problem _____

MODEL Make simplifying assumptions.

VISUALIZE

- Draw a picture. Show important points in the motion.
- Establish a coordinate system. Define symbols.
- List knowns. Identify what you're trying to find.
- Draw a motion diagram.
- Identify forces and interactions.
- Draw free-body diagrams.

Known

Find

SOLVE

Start with Newton's first or second law in component form, adding other information as needed to solve the problem.

ASSESS

Have you answered the question?
Do you have correct units, signs, and significant figures?
Is your answer reasonable?

DYNAMICS WORKSHEET   Name _____   Problem _____

MODEL   Make simplifying assumptions.

VISUALIZE
- Draw a picture. Show important points in the motion.
- Establish a coordinate system. Define symbols.
- List knowns. Identify what you're trying to find.
- Draw a motion diagram.
- Identify forces and interactions.
- Draw free-body diagrams.

Known

Find

SOLVE
Start with Newton's first or second law in component form, adding other information as needed to solve the problem.

ASSESS
Have you answered the question?
Do you have correct units, signs, and significant figures?
Is your answer reasonable?

DYNAMICS WORKSHEET   Name _____ Problem _____

MODEL Make simplifying assumptions.

VISUALIZE

- Draw a picture. Show important points in the motion.
- Establish a coordinate system. Define symbols.
- List knowns. Identify what you're trying to find.

- Draw a motion diagram.
- Identify forces and interactions.
- Draw free-body diagrams.

Known

Find

SOLVE
Start with Newton's first or second law in component form, adding other information as needed to solve the problem.

ASSESS
Have you answered the question?
Do you have correct units, signs, and significant figures?
Is your answer reasonable?

DYNAMICS WORKSHEET   Name _____   Problem _____

MODEL **MODEL** Make simplifying assumptions.

VISUALIZE

- Draw a picture. Show important points in the motion.
- Establish a coordinate system. Define symbols.
- List knowns. Identify what you're trying to find.

- Draw a motion diagram.
- Identify forces and interactions.
- Draw free-body diagrams.

Known

Find

SOLVE

Start with Newton's first or second law in component form, adding other information as needed to solve the problem.

ASSESS

Have you answered the question?

Do you have correct units, signs, and significant figures?

Is your answer reasonable?

DYNAMICS WORKSHEET Name _____ Problem _____

MODEL Make simplifying assumptions.

VISUALIZE

- Draw a picture. Show important points in the motion.
- Establish a coordinate system. Define symbols.
- List knowns. Identify what you're trying to find.

- Draw a motion diagram.
- Identify forces and interactions.
- Draw free-body diagrams.

Known

Find

SOLVE
Start with Newton's first or second law in component form, adding other information as needed to solve the problem.

ASSESS
Have you answered the question?
Do you have correct units, signs, and significant figures?
Is your answer reasonable?

DYNAMICS WORKSHEET Name _____ Problem _____

MODEL Make simplifying assumptions.

VISUALIZE

- Draw a picture. Show important points in the motion.
- Establish a coordinate system. Define symbols.
- List knowns. Identify what you're trying to find.

- Draw a motion diagram.
- Identify forces and interactions.
- Draw free-body diagrams.

Known

Find

SOLVE

Start with Newton's first or second law in component form, adding other information as needed to solve the problem.

ASSESS

Have you answered the question?

Do you have correct units, signs, and significant figures?

Is your answer reasonable?

DYNAMICS WORKSHEET   Name _____   Problem _____

MODEL Make simplifying assumptions.

VISUALIZE

- Draw a picture. Show important points in the motion.
- Establish a coordinate system. Define symbols.
- List knowns. Identify what you're trying to find.

- Draw a motion diagram.
- Identify forces and interactions.
- Draw free-body diagrams.

Known

Find

SOLVE
Start with Newton's first or second law in component form, adding other information as needed to solve the problem.

ASSESS
Have you answered the question?
Do you have correct units, signs, and significant figures?
Is your answer reasonable?

DYNAMICS WORKSHEET   Name _____   Problem _____

MODEL   Make simplifying assumptions.

VISUALIZE

- Draw a picture. Show important points in the motion.
- Establish a coordinate system. Define symbols.
- List knowns. Identify what you're trying to find.

- Draw a motion diagram.
- Identify forces and interactions.
- Draw free-body diagrams.

Known

Find

SOLVE

Start with Newton's first or second law in component form, adding other information as needed to solve the problem.

ASSESS

Have you answered the question?

Do you have correct units, signs, and significant figures?

Is your answer reasonable?

DYNAMICS WORKSHEET   Name _____   Problem _____

MODEL Make simplifying assumptions.

VISUALIZE
- Draw a picture. Show important points in the motion.
- Establish a coordinate system. Define symbols.
- List knowns. Identify what you're trying to find.
- Draw a motion diagram.
- Identify forces and interactions.
- Draw free-body diagrams.

Known

Find

SOLVE
Start with Newton's first or second law in component form, adding other information as needed to solve the problem.

ASSESS
Have you answered the question?
Do you have correct units, signs, and significant figures?
Is your answer reasonable?

DYNAMICS WORKSHEET   Name _____   Problem _____

MODEL Make simplifying assumptions.

VISUALIZE

- Draw a picture. Show important points in the motion.
- Establish a coordinate system. Define symbols.
- List knowns. Identify what you're trying to find.

- Draw a motion diagram.
- Identify forces and interactions.
- Draw free-body diagrams.

Known

Find

SOLVE

Start with Newton's first or second law in component form, adding other information as needed to solve the problem.

DYNAMICS WORKSHEET Name _____ Problem _____

MODEL Make simplifying assumptions.

VISUALIZE
- Draw a picture. Show important points in the motion.
- Establish a coordinate system. Define symbols.
- List knowns. Identify what you're trying to find.
- Draw a motion diagram.
- Identify forces and interactions.
- Draw free-body diagrams.

Known

Find

SOLVE
Start with Newton's first or second law in component form, adding other information as needed to solve the problem.

ASSESS
Have you answered the question?
Do you have correct units, signs, and significant figures?
Is your answer reasonable?

DYNAMICS WORKSHEET   Name _____   Problem _____

MODEL Make simplifying assumptions.

VISUALIZE

- Draw a picture. Show important points in the motion.
- Establish a coordinate system. Define symbols.
- List knowns. Identify what you're trying to find.

- Draw a motion diagram.
- Identify forces and interactions.
- Draw free-body diagrams.

Known

Find

SOLVE
Start with Newton's first or second law in component form, adding other information as needed to solve the problem.

ASSESS
Have you answered the question?
Do you have correct units, signs, and significant figures?
Is your answer reasonable?

DYNAMICS WORKSHEET Name _____ Problem _____

MODEL Make simplifying assumptions.

VISUALIZE

- Draw a picture. Show important points in the motion.
- Establish a coordinate system. Define symbols.
- List knowns. Identify what you're trying to find.
- Draw a motion diagram.
- Identify forces and interactions.
- Draw free-body diagrams.

Known

Find

SOLVE

Start with Newton's first or second law in component form, adding other information as needed to solve the problem.

ASSESS

Have you answered the question?

Do you have correct units, signs, and significant figures?

Is your answer reasonable?

DYNAMICS WORKSHEET   Name _____   Problem _____

MODEL Make simplifying assumptions.

VISUALIZE

- Draw a picture. Show important points in the motion.
- Establish a coordinate system. Define symbols.
- List knowns. Identify what you're trying to find.
- Draw a motion diagram.
- Identify forces and interactions.
- Draw free-body diagrams.

Known

Find

SOLVE

Start with Newton's first or second law in component form, adding other information as needed to solve the problem.

ASSESS

Have you answered the question?

Do you have correct units, signs, and significant figures?

Is your answer reasonable?

DYNAMICS WORKSHEET   Name _____   Problem _____

MODEL Make simplifying assumptions.

VISUALIZE

- Draw a picture. Show important points in the motion.
- Establish a coordinate system. Define symbols.
- List knowns. Identify what you're trying to find.

- Draw a motion diagram.
- Identify forces and interactions.
- Draw free-body diagrams.

Known

Find

SOLVE

Start with Newton's first or second law in component form, adding other information as needed to solve the problem.

ASSESS

Have you answered the question?

Do you have correct units, signs, and significant figures?

Is your answer reasonable?

DYNAMICS WORKSHEET   Name _____ Problem _____

MODEL   Make simplifying assumptions.

VISUALIZE

- Draw a picture. Show important points in the motion.
- Establish a coordinate system. Define symbols.
- List knowns. Identify what you're trying to find.
- Draw a motion diagram.
- Identify forces and interactions.
- Draw free-body diagrams.

Known

Find

SOLVE

Start with Newton's first or second law in component form, adding other information as needed to solve the problem.

ASSESS

Have you answered the question?

Do you have correct units, signs, and significant figures?

Is your answer reasonable?

DYNAMICS WORKSHEET Name _____ Problem _____

MODEL Make simplifying assumptions.

VISUALIZE

- Draw a picture. Show important points in the motion.
- Establish a coordinate system. Define symbols.
- List knowns. Identify what you're trying to find.
- Draw a motion diagram.
- Identify forces and interactions.
- Draw free-body diagrams.

Known

Find

SOLVE
Start with Newton's first or second law in component form, adding other information as needed to solve the problem.

ASSESS
Have you answered the question?
Do you have correct units, signs, and significant figures?
Is your answer reasonable?

DYNAMICS WORKSHEET   Name _____   Problem _____

MODEL Make simplifying assumptions.

VISUALIZE

- Draw a picture. Show important points in the motion.
- Establish a coordinate system. Define symbols.
- List knowns. Identify what you're trying to find.
- Draw a motion diagram.
- Identify forces and interactions.
- Draw free-body diagrams.

Known

Find

SOLVE

Start with Newton's first or second law in component form, adding other information as needed to solve the problem.

ASSESS

Have you answered the question?

Do you have correct units, signs, and significant figures?

Is your answer reasonable?

DYNAMICS WORKSHEET   Name _____   Problem _____

MODEL   Make simplifying assumptions.

VISUALIZE
- Draw a picture. Show important points in the motion.
- Establish a coordinate system. Define symbols.
- List knowns. Identify what you're trying to find.

- Draw a motion diagram.
- Identify forces and interactions.
- Draw free-body diagrams.

Known

Find

SOLVE
Start with Newton's first or second law in component form, adding other information as needed to solve the problem.

ASSESS
Have you answered the question?
Do you have correct units, signs, and significant figures?
Is your answer reasonable?

DYNAMICS WORKSHEET   Name _____   Problem _____

MODEL   Make simplifying assumptions.

VISUALIZE

- Draw a picture. Show important points in the motion.
- Establish a coordinate system. Define symbols.
- List knowns. Identify what you're trying to find.
- Draw a motion diagram.
- Identify forces and interactions.
- Draw free-body diagrams.

Known

Find

SOLVE
Start with Newton's first or second law in component form, adding other information as needed to solve the problem.

ASSESS
Have you answered the question?
Do you have correct units, signs, and significant figures?
Is your answer reasonable?

DYNAMICS WORKSHEET   Name _____   Problem _____

MODEL   Make simplifying assumptions.

VISUALIZE

- Draw a picture. Show important points in the motion.
- Establish a coordinate system. Define symbols.
- List knowns. Identify what you're trying to find.

- Draw a motion diagram.
- Identify forces and interactions.
- Draw free-body diagrams.

Known

Find

SOLVE

Start with Newton's first or second law in component form, adding other information as needed to solve the problem.

ASSESS

Have you answered the question?
Do you have correct units, signs, and significant figures?
Is your answer reasonable?

DYNAMICS WORKSHEET   Name _____   Problem _____

MODEL Make simplifying assumptions.

VISUALIZE

- Draw a picture. Show important points in the motion.
- Establish a coordinate system. Define symbols.
- List knowns. Identify what you're trying to find.
- Draw a motion diagram.
- Identify forces and interactions.
- Draw free-body diagrams.

Known

Find

SOLVE

Start with Newton's first or second law in component form, adding other information as needed to solve the problem.

ASSESS

Have you answered the question?
Do you have correct units, signs, and significant figures?
Is your answer reasonable?

DYNAMICS WORKSHEET   Name _____   Problem _____

VISUALIZE

- Draw a picture. Show important points in the motion.
- Establish a coordinate system. Define symbols.
- List knowns. Identify what you're trying to find.

- Draw a motion diagram.
- Identify forces and interactions.
- Draw free-body diagrams.

Known

Find

SOLVE

Start with Newton's first or second law in component form, adding other information as needed to solve the problem.

ASSESS

Have you answered the question?
Do you have correct units, signs, and significant figures?
Is your answer reasonable?

DYNAMICS WORKSHEET   Name _____   Problem _____

MODEL Make simplifying assumptions.

VISUALIZE

- Draw a picture. Show important points in the motion.
- Establish a coordinate system. Define symbols.
- List knowns. Identify what you're trying to find.
- Draw a motion diagram.
- Identify forces and interactions.
- Draw free-body diagrams.

Known

Find

SOLVE

Start with Newton's first or second law in component form, adding other information as needed to solve the problem.

ASSESS

Have you answered the question?

Do you have correct units, signs, and significant figures?

Is your answer reasonable?

DYNAMICS WORKSHEET Name _____ Problem _____

MODEL Make simplifying assumptions.

VISUALIZE

- Draw a picture. Show important points in the motion.
- Establish a coordinate system. Define symbols.
- List knowns. Identify what you're trying to find.
- Draw a motion diagram.
- Identify forces and interactions.
- Draw free-body diagrams.

Known

Find

SOLVE

Start with Newton's first or second law in component form, adding other information as needed to solve the problem.

ASSESS

Have you answered the question?

Do you have correct units, signs, and significant figures?

Is your answer reasonable?

DYNAMICS WORKSHEET   Name _____   Problem _____

VISUALIZE

- Draw a picture. Show important points in the motion.
- Establish a coordinate system. Define symbols.
- List knowns. Identify what you're trying to find.
- Draw a motion diagram.
- Identify forces and interactions.
- Draw free-body diagrams.

Known

Find

SOLVE

Start with Newton's first or second law in component form, adding other information as needed to solve the problem.

ASSESS

Have you answered the question?

Do you have correct units, signs, and significant figures?

Is your answer reasonable?

DYNAMICS WORKSHEET   Name _____   Problem _____

MODEL Make simplifying assumptions.

VISUALIZE

- Draw a picture. Show important points in the motion.
- Establish a coordinate system. Define symbols.
- List knowns. Identify what you're trying to find.
- Draw a motion diagram.
- Identify forces and interactions.
- Draw free-body diagrams.

Known

Find

SOLVE

Start with Newton's first or second law in component form, adding other information as needed to solve the problem.

ASSESS

Have you answered the question?

Do you have correct units, signs, and significant figures?

Is your answer reasonable?

DYNAMICS WORKSHEET Name _____ Problem _____

MODEL Make simplifying assumptions.

VISUALIZE
- Draw a picture. Show important points in the motion.
- Establish a coordinate system. Define symbols.
- List knowns. Identify what you're trying to find.

- Draw a motion diagram.
- Identify forces and interactions.
- Draw free-body diagrams.

Known

Find

SOLVE
Start with Newton's first or second law in component form, adding other information as needed to solve the problem.

ASSESS
Have you answered the question?
Do you have correct units, signs, and significant figures?
Is your answer reasonable?

DYNAMICS WORKSHEET Name _____ Problem _____

MODEL Make simplifying assumptions.

VISUALIZE
- Draw a picture. Show important points in the motion.
- Establish a coordinate system. Define symbols.
- List knowns. Identify what you're trying to find.
- Draw a motion diagram.
- Identify forces and interactions.
- Draw free-body diagrams.

Known

Find

SOLVE
Start with Newton's first or second law in component form, adding other information as needed to solve the problem.

ASSESS
Have you answered the question?
Do you have correct units, signs, and significant figures?
Is your answer reasonable?

DYNAMICS WORKSHEET   Name _____   Problem _____

MODEL Make simplifying assumptions.

VISUALIZE
- Draw a picture. Show important points in the motion.
- Establish a coordinate system. Define symbols.
- List knowns. Identify what you're trying to find.
- Draw a motion diagram.
- Identify forces and interactions.
- Draw free-body diagrams.

Known

Find

SOLVE
Start with Newton's first or second law in component form, adding other information as needed to solve the problem.

ASSESS
Have you answered the question?
Do you have correct units, signs, and significant figures?
Is your answer reasonable?

DYNAMICS WORKSHEET Name _____ Problem _____

VISUALIZE

- Draw a picture. Show important points in the motion.
- Establish a coordinate system. Define symbols.
- List knowns. Identify what you're trying to find.

- Draw a motion diagram.
- Identify forces and interactions.
- Draw free-body diagrams.

Known

Find

SOLVE

Start with Newton's first or second law in component form, adding other information as needed to solve the problem.

ASSESS

Have you answered the question?
Do you have correct units, signs, and significant figures?
Is your answer reasonable?

DYNAMICS WORKSHEET   Name _____   Problem _____

MODEL Make simplifying assumptions.

VISUALIZE

- Draw a picture. Show important points in the motion.
- Establish a coordinate system. Define symbols.
- List knowns. Identify what you're trying to find.
- Draw a motion diagram.
- Identify forces and interactions.
- Draw free-body diagrams.

Known

Find

SOLVE
Start with Newton's first or second law in component form, adding other information as needed to solve the problem.

ASSESS
Have you answered the question?
Do you have correct units, signs, and significant figures?
Is your answer reasonable?

MOMENTUM WORKSHEET   Name _____   Problem _____

MODEL Make simplifying assumptions.

VISUALIZE

- Draw a before-and-after picture.
- Establish a coordinate system. Define symbols.

- Draw a momentum bar chart.
- List knowns. Identify what you're trying to find.

Known

Find

- What is the system? _____
- What forces exert impulses on the system? _____
- Is the system's momentum conserved during part or all of the problem?

  If so, during which part? _____

  _____

$$P_{ix} \quad + \quad J_x \quad = \quad P_{fx}$$

SOLVE

Start with conservation of momentum or the impulse-momentum theorem, using Newton's laws or kinematics as needed.

ASSESS

Have you answered the question?
Do you have correct units, signs, and significant figures?
Is your answer reasonable?

MOMENTUM WORKSHEET   Name _____   Problem _____

MODEL   Make simplifying assumptions.

VISUALIZE

- Draw a before-and-after picture.
- Establish a coordinate system. Define symbols.

- Draw a momentum bar chart.
- List knowns. Identify what you're trying to find.

Known

Find

- What is the system? _____
- What forces exert impulses on the system? _____
- Is the system's momentum conserved during part or all of the problem?

  If so, during which part? _____

  _____

$$+$$

$$0$$

$$-$$

$$P_{ix} \quad + \quad J_x \quad = \quad P_{fx}$$

SOLVE

Start with conservation of momentum or the impulse-momentum theorem, using Newton's laws or kinematics as needed.

ASSESS

Have you answered the question?
Do you have correct units, signs, and significant figures?
Is your answer reasonable?

MOMENTUM WORKSHEET Name _____ Problem _____

MODEL Make simplifying assumptions.

VISUALIZE

- Draw a before-and-after picture.
- Establish a coordinate system. Define symbols.

- Draw a momentum bar chart.
- List knowns. Identify what you're trying to find.

Known

Find

- What is the system? _____
- What forces exert impulses on the system? _____
- Is the system's momentum conserved during part or all of the problem?

  If so, during which part? _____

  _____

$$+$$

$$0$$

$$-$$

$$P_{ix} \quad + \quad J_x \quad = \quad P_{fx}$$

SOLVE

Start with conservation of momentum or the impulse-momentum theorem, using Newton's laws or kinematics as needed.

ASSESS

Have you answered the question?
Do you have correct units, signs, and significant figures?
Is your answer reasonable?

MOMENTUM WORKSHEET   Name _____   Problem _____

MODEL Make simplifying assumptions.

VISUALIZE
- Draw a before-and-after picture.
- Establish a coordinate system. Define symbols.

- Draw a momentum bar chart.
- List knowns. Identify what you're trying to find.

Known

Find

- What is the system? _____
- What forces exert impulses on the system? _____
- Is the system's momentum conserved during part or all of the problem?

  If so, during which part? _____

  _____

$$+ \quad 0 \quad -$$

$$P_{ix} \quad + \quad J_x \quad = \quad P_{fx}$$

SOLVE
Start with conservation of momentum or the impulse-momentum theorem, using Newton's laws or kinematics as needed.

ASSESS
Have you answered the question?
Do you have correct units, signs, and significant figures?
Is your answer reasonable?

MOMENTUM WORKSHEET   Name _____   Problem _____

MODEL Make simplifying assumptions.

VISUALIZE

- Draw a before-and-after picture.
- Establish a coordinate system. Define symbols.

- Draw a momentum bar chart.
- List knowns. Identify what you're trying to find.

Known

Find

- What is the system? _____
- What forces exert impulses on the system? _____
- Is the system's momentum conserved during part or all of the problem?

  If so, during which part? _____

  _____

$$P_{ix} \quad + \quad J_x \quad = \quad P_{fx}$$

SOLVE

Start with conservation of momentum or the impulse-momentum theorem, using Newton's laws or kinematics as needed.

ASSESS

Have you answered the question?
Do you have correct units, signs, and significant figures?
Is your answer reasonable?

MOMENTUM WORKSHEET   Name _____   Problem _____

VISUALIZE

- Draw a before-and-after picture.
- Establish a coordinate system. Define symbols.
- Draw a momentum bar chart.
- List knowns. Identify what you're trying to find.

Known

Find

- What is the system? _____
- What forces exert impulses on the system? _____
- Is the system's momentum conserved during part or all of the problem?

  If so, during which part? _____

  _____

$$+$$

$$0$$

$$-$$

$$P_{ix} \quad + \quad J_x \quad = \quad P_{fx}$$

SOLVE

Start with conservation of momentum or the impulse-momentum theorem, using Newton's laws or kinematics as needed.

ASSESS

Have you answered the question?
Do you have correct units, signs, and significant figures?
Is your answer reasonable?

MOMENTUM WORKSHEET Name _____ Problem _____

MODEL Make simplifying assumptions.

VISUALIZE

- Draw a before-and-after picture.
- Establish a coordinate system. Define symbols.

- Draw a momentum bar chart.
- List knowns. Identify what you're trying to find.

Known

Find

- What is the system? _____
- What forces exert impulses on the system? _____
- Is the system's momentum conserved during part or all of the problem?

  If so, during which part? _____

  _____

$$P_{ix} + J_x = P_{fx}$$

SOLVE

Start with conservation of momentum or the impulse-momentum theorem, using Newton's laws or kinematics as needed.

ASSESS

Have you answered the question?
Do you have correct units, signs, and significant figures?
Is your answer reasonable?

MOMENTUM WORKSHEET   Name _____   Problem _____

MODEL Make simplifying assumptions.

_____

VISUALIZE

- Draw a before-and-after picture.
- Establish a coordinate system. Define symbols.

- Draw a momentum bar chart.
- List knowns. Identify what you're trying to find.

Known

Find

- What is the system? _____
- What forces exert impulses on the system? _____
- Is the system's momentum conserved during part or all of the problem?

   If so, during which part? _____

   _____

$$+$$

$$0 \qquad \underline{\quad} \quad + \quad \underline{\quad} \quad = \quad \underline{\quad}$$

$$-$$

$$P_{ix} \quad + \quad J_x \quad = \quad P_{fx}$$

SOLVE

Start with conservation of momentum or the impulse-momentum theorem, using Newton's laws or kinematics as needed.

ASSESS

Have you answered the question?
Do you have correct units, signs, and significant figures?
Is your answer reasonable?

MOMENTUM WORKSHEET   Name _____   Problem _____

MODEL Make simplifying assumptions.

VISUALIZE

- Draw a before-and-after picture.
- Establish a coordinate system. Define symbols.

- Draw a momentum bar chart.
- List knowns. Identify what you're trying to find.

Known

Find

- What is the system? _____
- What forces exert impulses on the system? _____
- Is the system's momentum conserved during part or all of the problem?

  If so, during which part? _____

  _____

$$P_{ix} \quad + \quad J_x \quad = \quad P_{fx}$$

SOLVE

Start with conservation of momentum or the impulse-momentum theorem, using Newton's laws or kinematics as needed.

ASSESS

Have you answered the question?
Do you have correct units, signs, and significant figures?
Is your answer reasonable?

MOMENTUM WORKSHEET   Name _____   Problem _____

MODEL Make simplifying assumptions.

VISUALIZE

- Draw a before-and-after picture.
- Establish a coordinate system. Define symbols.

- Draw a momentum bar chart.
- List knowns. Identify what you're trying to find.

Known

Find

- What is the system? _____
- What forces exert impulses on the system? _____
- Is the system's momentum conserved during part or all of the problem?

  If so, during which part? _____

  _____

$$+ \qquad 0 \qquad -$$

$$P_{ix} \quad + \quad J_x \quad = \quad P_{fx}$$

SOLVE

Start with conservation of momentum or the impulse-momentum theorem, using Newton's laws or kinematics as needed.

ASSESS

Have you answered the question?
Do you have correct units, signs, and significant figures?
Is your answer reasonable?

MOMENTUM WORKSHEET Name _____ Problem _____

MODEL Make simplifying assumptions.

VISUALIZE

- Draw a before-and-after picture.
- Establish a coordinate system. Define symbols.

- Draw a momentum bar chart.
- List knowns. Identify what you're trying to find.

Known

Find

- What is the system? _____
- What forces exert impulses on the system? _____
- Is the system's momentum conserved during part or all of the problem?

  If so, during which part? _____

  _____

$$P_{ix} \quad + \quad J_x \quad = \quad P_{fx}$$

SOLVE

Start with conservation of momentum or the impulse-momentum theorem, using Newton's laws or kinematics as needed.

ASSESS

Have you answered the question?
Do you have correct units, signs, and significant figures?
Is your answer reasonable?

MOMENTUM WORKSHEET   Name _____   Problem _____

MODEL   Make simplifying assumptions.

VISUALIZE

- Draw a before-and-after picture.
- Establish a coordinate system. Define symbols.
- Draw a momentum bar chart.
- List knowns. Identify what you're trying to find.

Known

Find

- What is the system? _____
- What forces exert impulses on the system? _____
- Is the system's momentum conserved during part or all of the problem?

  If so, during which part? _____

  _____

$$+$$

$$0$$

$$-$$

$$P_{ix} \quad + \quad J_x \quad = \quad P_{fx}$$

SOLVE

Start with conservation of momentum or the impulse-momentum theorem, using Newton's laws or kinematics as needed.

ASSESS

Have you answered the question?
Do you have correct units, signs, and significant figures?
Is your answer reasonable?

ENERGY WORKSHEET

Name _____ Problem _____

MODEL Make simplifying assumptions.

VISUALIZE

- Draw a before-and-after picture.
- Establish a coordinate system. Define symbols.

- Draw an energy bar chart.
- List knowns. Identify what you're trying to find.

Known

Find

What is the system? _____

Potential energies? _____

Nonconservative forces? _____

External forces? _____

Is mechanical energy conserved? _____

$$K_i + U_i + W_{ext} = K_f + U_f + \Delta E_{th}$$

SOLVE

Start with conservation of energy, adding other information and techniques as needed to solve the problem.

ASSESS

Have you answered the question?

Do you have correct units, signs, and significant figures?

Is your answer reasonable?

ENERGY WORKSHEET

Name _____ Problem _____

MODEL Make simplifying assumptions.

VISUALIZE

- Draw a before-and-after picture.
- Establish a coordinate system. Define symbols.
- Draw an energy bar chart.
- List knowns. Identify what you're trying to find.

Known

Find

What is the system? _____

Potential energies? _____

Nonconservative forces? _____

External forces? _____

Is mechanical energy conserved? _____

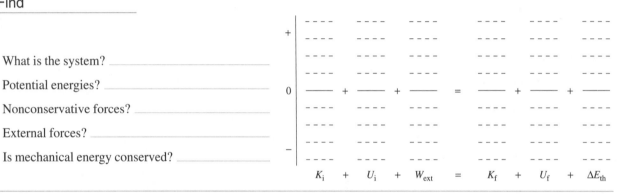

SOLVE

Start with conservation of energy, adding other information and techniques as needed to solve the problem.

ASSESS

Have you answered the question?
Do you have correct units, signs, and significant figures?
Is your answer reasonable?

ENERGY WORKSHEET

Name _____ Problem _____

MODEL Make simplifying assumptions.

VISUALIZE

- Draw a before-and-after picture.
- Establish a coordinate system. Define symbols.

- Draw an energy bar chart.
- List knowns. Identify what you're trying to find.

Known

Find

What is the system? _____

Potential energies? _____

Nonconservative forces? _____

External forces? _____

Is mechanical energy conserved? _____

$$K_i \quad + \quad U_i \quad + \quad W_{ext} \quad = \quad K_f \quad + \quad U_f \quad + \quad \Delta E_{th}$$

SOLVE

Start with conservation of energy, adding other information and techniques as needed to solve the problem.

ASSESS

Have you answered the question?

Do you have correct units, signs, and significant figures?

Is your answer reasonable?

ENERGY WORKSHEET

Name _____ Problem _____

MODEL Make simplifying assumptions.

VISUALIZE

- Draw a before-and-after picture.
- Establish a coordinate system. Define symbols.

- Draw an energy bar chart.
- List knowns. Identify what you're trying to find.

Known

Find

What is the system? _____

Potential energies? _____

Nonconservative forces? _____

External forces? _____

Is mechanical energy conserved? _____

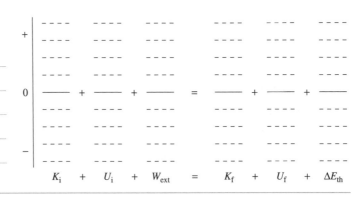

$$K_i \ + \ U_i \ + \ W_{ext} \ = \ K_f \ + \ U_f \ + \ \Delta E_{th}$$

SOLVE

Start with conservation of energy, adding other information and techniques as needed to solve the problem.

ASSESS

Have you answered the question?

Do you have correct units, signs, and significant figures?

Is your answer reasonable?

ENERGY WORKSHEET

Name _____ Problem _____

MODEL Make simplifying assumptions.

VISUALIZE

- Draw a before-and-after picture.
- Establish a coordinate system. Define symbols.

- Draw an energy bar chart.
- List knowns. Identify what you're trying to find.

Known

Find

What is the system? _____

Potential energies? _____

Nonconservative forces? _____

External forces? _____

Is mechanical energy conserved? _____

$$K_i \quad + \quad U_i \quad + \quad W_{ext} \quad = \quad K_f \quad + \quad U_f \quad + \quad \Delta E_{th}$$

SOLVE

Start with conservation of energy, adding other information and techniques as needed to solve the problem.

ASSESS

Have you answered the question?
Do you have correct units, signs, and significant figures?
Is your answer reasonable?

ENERGY WORKSHEET   Name _____ Problem _____

MODEL Make simplifying assumptions.

VISUALIZE
- Draw a before-and-after picture.
- Establish a coordinate system. Define symbols.
- Draw an energy bar chart.
- List knowns. Identify what you're trying to find.

Known

Find

What is the system? _____

Potential energies? _____

Nonconservative forces? _____

External forces? _____

Is mechanical energy conserved? _____

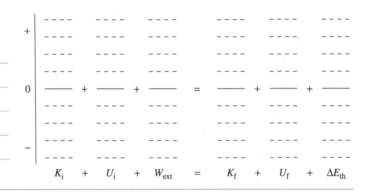

$$K_i \;+\; U_i \;+\; W_{ext} \;=\; K_f \;+\; U_f \;+\; \Delta E_{th}$$

SOLVE
Start with conservation of energy, adding other information and techniques as needed to solve the problem.

ASSESS
Have you answered the question?
Do you have correct units, signs, and significant figures?
Is your answer reasonable?

ENERGY WORKSHEET

Name _____ Problem _____

MODEL Make simplifying assumptions.

VISUALIZE

- Draw a before-and-after picture.
- Establish a coordinate system. Define symbols.

- Draw an energy bar chart.
- List knowns. Identify what you're trying to find.

Known

Find

What is the system? _____

Potential energies? _____

Nonconservative forces? _____

External forces? _____

Is mechanical energy conserved? _____

$$K_i + U_i + W_{ext} = K_f + U_f + \Delta E_{th}$$

SOLVE

Start with conservation of energy, adding other information and techniques as needed to solve the problem.

ASSESS

Have you answered the question?
Do you have correct units, signs, and significant figures?
Is your answer reasonable?

ENERGY WORKSHEET   Name _____   Problem _____

MODEL Make simplifying assumptions.

VISUALIZE

- Draw a before-and-after picture.
- Establish a coordinate system. Define symbols.

- Draw an energy bar chart.
- List knowns. Identify what you're trying to find.

Known

Find

What is the system? _____

Potential energies? _____

Nonconservative forces? _____

External forces? _____

Is mechanical energy conserved? _____

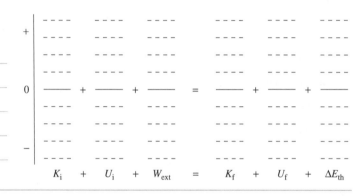

$$K_i \quad + \quad U_i \quad + \quad W_{ext} \quad = \quad K_f \quad + \quad U_f \quad + \quad \Delta E_{th}$$

SOLVE

Start with conservation of energy, adding other information and techniques as needed to solve the problem.

ASSESS

Have you answered the question?

Do you have correct units, signs, and significant figures?

Is your answer reasonable?

ENERGY WORKSHEET   Name _____   Problem _____

MODEL Make simplifying assumptions.

VISUALIZE
- Draw a before-and-after picture.
- Establish a coordinate system. Define symbols.
- Draw an energy bar chart.
- List knowns. Identify what you're trying to find.

Known

Find

What is the system? _____

Potential energies? _____

Nonconservative forces? _____

External forces? _____

Is mechanical energy conserved? _____

$$K_i + U_i + W_{ext} = K_f + U_f + \Delta E_{th}$$

SOLVE

Start with conservation of energy, adding other information and techniques as needed to solve the problem.

ASSESS
Have you answered the question?
Do you have correct units, signs, and significant figures?
Is your answer reasonable?

ENERGY WORKSHEET

Name _____ Problem _____

MODEL Make simplifying assumptions.

VISUALIZE

- Draw a before-and-after picture.
- Establish a coordinate system. Define symbols.
- Draw an energy bar chart.
- List knowns. Identify what you're trying to find.

Known

Find

What is the system? _____

Potential energies? _____

Nonconservative forces? _____

External forces? _____

Is mechanical energy conserved? _____

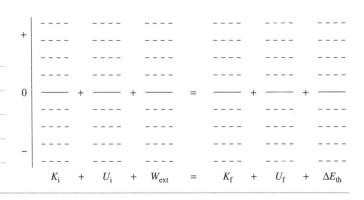

$$K_i \;+\; U_i \;+\; W_{ext} \;=\; K_f \;+\; U_f \;+\; \Delta E_{th}$$

SOLVE

Start with conservation of energy, adding other information and techniques as needed to solve the problem.

ASSESS

Have you answered the question?
Do you have correct units, signs, and significant figures?
Is your answer reasonable?

ENERGY WORKSHEET   Name _____ Problem _____

VISUALIZE

- Draw a before-and-after picture.
- Establish a coordinate system. Define symbols.

- Draw an energy bar chart.
- List knowns. Identify what you're trying to find.

Known

Find

What is the system? _____

Potential energies? _____

Nonconservative forces? _____

External forces? _____

Is mechanical energy conserved? _____

$$K_i \quad + \quad U_i \quad + \quad W_{ext} \quad = \quad K_f \quad + \quad U_f \quad + \quad \Delta E_{th}$$

SOLVE

Start with conservation of energy, adding other information and techniques as needed to solve the problem.

ASSESS

Have you answered the question?

Do you have correct units, signs, and significant figures?

Is your answer reasonable?

ENERGY WORKSHEET

Name _____ Problem _____

MODEL Make simplifying assumptions.

VISUALIZE

- Draw a before-and-after picture.
- Establish a coordinate system. Define symbols.

- Draw an energy bar chart.
- List knowns. Identify what you're trying to find.

Known

Find

What is the system? _____

Potential energies? _____

Nonconservative forces? _____

External forces? _____

Is mechanical energy conserved? _____

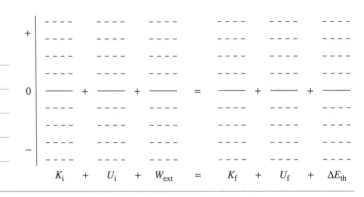

SOLVE

Start with conservation of energy, adding other information and techniques as needed to solve the problem.

ASSESS

Have you answered the question?
Do you have correct units, signs, and significant figures?
Is your answer reasonable?

ENERGY WORKSHEET   Name _____ Problem _____

MODEL Make simplifying assumptions.

VISUALIZE

- Draw a before-and-after picture.
- Establish a coordinate system. Define symbols.

- Draw an energy bar chart.
- List knowns. Identify what you're trying to find.

Known

Find

What is the system? _____

Potential energies? _____

Nonconservative forces? _____

External forces? _____

Is mechanical energy conserved? _____

$$K_i + U_i + W_{ext} = K_f + U_f + \Delta E_{th}$$

SOLVE

Start with conservation of energy, adding other information and techniques as needed to solve the problem.

ASSESS

Have you answered the question?
Do you have correct units, signs, and significant figures?
Is your answer reasonable?

ENERGY WORKSHEET

Name _____ Problem _____

MODEL Make simplifying assumptions.

VISUALIZE

- Draw a before-and-after picture.
- Establish a coordinate system. Define symbols.

- Draw an energy bar chart.
- List knowns. Identify what you're trying to find.

Known

Find

What is the system? _____

Potential energies? _____

Nonconservative forces? _____

External forces? _____

Is mechanical energy conserved? _____

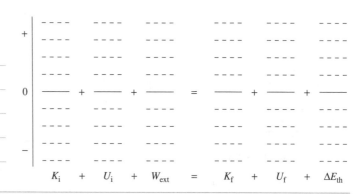

SOLVE

Start with conservation of energy, adding other information and techniques as needed to solve the problem.

ASSESS

Have you answered the question?

Do you have correct units, signs, and significant figures?

Is your answer reasonable?

ENERGY WORKSHEET

Name _____ Problem _____

MODEL Make simplifying assumptions.

VISUALIZE

- Draw a before-and-after picture.
- Establish a coordinate system. Define symbols.

- Draw an energy bar chart.
- List knowns. Identify what you're trying to find.

Known

Find

What is the system? _____

Potential energies? _____

Nonconservative forces? _____

External forces? _____

Is mechanical energy conserved? _____

$$K_i \ + \ U_i \ + \ W_{ext} \ = \ K_f \ + \ U_f \ + \ \Delta E_{th}$$

SOLVE

Start with conservation of energy, adding other information and techniques as needed to solve the problem.

ASSESS

Have you answered the question?

Do you have correct units, signs, and significant figures?

Is your answer reasonable?

ENERGY WORKSHEET

Name _____ Problem _____

MODEL Make simplifying assumptions.

VISUALIZE
- Draw a before-and-after picture.
- Establish a coordinate system. Define symbols.
- Draw an energy bar chart.
- List knowns. Identify what you're trying to find.

Known

Find

What is the system? _____

Potential energies? _____

Nonconservative forces? _____

External forces? _____

Is mechanical energy conserved? _____

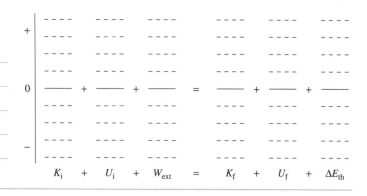

$$K_i \;+\; U_i \;+\; W_{ext} \;=\; K_f \;+\; U_f \;+\; \Delta E_{th}$$

SOLVE
Start with conservation of energy, adding other information and techniques as needed to solve the problem.

ASSESS
Have you answered the question?
Do you have correct units, signs, and significant figures?
Is your answer reasonable?